21世纪高等学校计算机类专业
核心课程系列教材

Android Studio
应用程序设计 第3版·微课视频版

◎ 张思民 编著

清华大学出版社
北京

内 容 简 介

本书是面向Android Studio初学者的入门教程，内容大致可以分成两个部分。第一部分（第1～4章）主要介绍Android Studio的安装、应用程序的结构、图形用户界面的组件及其设计方法；第二部分（第5～9章）主要介绍较高级的主题，内容包括后台服务与系统服务、网络通信、数据存储、应用Volley框架访问Web服务器、美颜与人脸检测等。

本书由浅入深、循序渐进地介绍Android Studio应用程序的设计方法和设计思想。本书讲解详细，内容丰富，每个知识点都配备了大量图示加以说明，并进行详细的分析解释，每章均精心选编了经典案例，对读者学习有很大的帮助，可以让读者轻松上手。本书提供了电子课件和所有例题的源代码，扫描每章提供的二维码可观看教学视频。

本书可作为高等院校及各类培训学校Android系统课程的教材，也可作为希望学习Android系统开发的读者自学用书。

本书封面贴有清华大学出版社防伪标签。无标签者不得销售。
版权所有，侵权必究。举报：010-62782989，beiqinquan@tup.tsinghua.edu.cn。

图书在版编目（CIP）数据

Android Studio应用程序设计：微课视频版/张思民编著. —3版. —北京：清华大学出版社，2023.3
（2024.2重印）
21世纪高等学校计算机类专业核心课程系列教材
ISBN 978-7-302-62979-5

Ⅰ. ①A… Ⅱ. ①张… Ⅲ. ①移动终端—应用程序—程序设计—高等学校—教材 Ⅳ. ①TN929.53

中国国家版本馆CIP数据核字（2023）第039743号

策划编辑：魏江江
责任编辑：王冰飞
封面设计：刘　键
责任校对：郝美丽
责任印制：刘海龙

出版发行：清华大学出版社
　　　　　网　　址：https://www.tup.com.cn，https://www.wqxuetang.con
　　　　　地　　址：北京清华大学学研大厦A座　　邮　　编：100084
　　　　　社 总 机：010-83470000　　邮　　购：010-62786544
　　　　　投稿与读者服务：010-62776969，c-service@tup.tsinghua.edu.cn
　　　　　质量反馈：010-62772015，zhiliang@tup.tsinghua.edu.cn
　　　　　课件下载：https://www.tup.com.cn，010-83470236
印 装 者：三河市人民印务有限公司
经　　销：全国新华书店
开　　本：185mm×260mm　　印　　张：16.5　　字　　数：402千字
版　　次：2013年3月第1版　2023年5月第3版　　印　　次：2024年2月第3次印刷
印　　数：35501～37500
定　　价：49.80元

产品编号：095900-01

第3版前言

　　党的二十大报告中指出：教育、科技、人才是全面建设社会主义现代化国家的基础性、战略性支撑。必须坚持科技是第一生产力、人才是第一资源、创新是第一动力，深入实施科教兴国战略、人才强国战略、创新驱动发展战略，这三大战略共同服务于创新型国家的建设。高等教育与经济社会发展紧密相连，对促进就业创业、助力经济社会发展、增进人民福祉具有重要意义。

　　由于 Android 系统发展很快，其版本一直在更新变化，Android 新版本对用户的操作权限进行了很多限制，从而导致旧版本的程序已经不能继续使用。与第 2 版相比，第 3 版的最大修改之处是根据 Android 11 重新改写了所有例题。

　　本书共 9 章。第 1 章主要讲解 Android Studio 开发环境的搭建，并介绍了开发 Android 应用程序的步骤和应用程序的结构；第 2~3 章讲解如何进行用户界面设计，介绍了用户图形界面的常用组件；第 4 章介绍图形与多媒体处理技术，介绍了绘制几何图形的基本方法、处理触摸屏事件的方法，还详细讨论了音频播放和视频播放的设计以及文本转换语音技术，并详细讲解了在 Android 中实现动画的技术；第 5 章介绍后台服务与系统服务；第 6 章介绍网络通信技术，介绍了 Web 视图以及基于 TCP 和基于 HTTP 的网络程序设计等网络编程技术；第 7 章介绍应用 Volley 框架访问 Web 服务器，并介绍了访问远程数据库的方法；第 8 章介绍数据存储技术，介绍了 SQLite 数据库存储方式、文件存储方式和 XML 文件的 SharedPreferences 存储方式；第 9 章讲解 OpenCV 应用实战——人脸美颜与人脸检测。

　　本书在例题代码中，为了区分系统自动生成的模板代码和其他部分的代码，所有系统生成的模板代码均以灰色底纹表示。对于不同的 Android 版本，系统生成的模板代码可能略有差异，一般不影响示例其他部分代码的正常运行。

　　为便于教学，本书提供丰富的配套资源，包括教学大纲、教学课件、电子教案、程序源码、习题答案和微课视频。

> ### 资源下载提示
>
> **课件等资源**：扫描封底的"课件下载"二维码，在公众号"书圈"下载。
> **素材（源码）等资源**：扫描目录上方的二维码下载。
> **视频等资源**：扫描封底的文泉云盘防盗码，再扫描书中相应章节中的二维码，可以在线学习。

　　由于计算机及软件技术发展迅速，加之作者水平有限，书中难免有不足和疏漏之处，希望广大读者与同行不吝赐教。

张思民

2023 年 3 月

目 录

资源下载

第 1 章 Android 系统及其开发过程 1

1.1 Android 系统概述 1
    1.1.1 Android 系统及其特点 1
    1.1.2 Android 的系统架构 2
    1.1.3 Android 开发分类 3

1.2 搭建 Android Studio 开发环境 🎥 4
    1.2.1 安装 Android Studio 前的必要准备 4
    1.2.2 安装 Android Studio 5

1.3 Android API 和在线帮助文档 7

1.4 Android 应用程序的开发过程 9
    1.4.1 开发 Android 应用程序的一般过程 9
    1.4.2 生成 Android 应用程序框架 9
    1.4.3 编写代码生成 MainActivity.java 11
    1.4.4 在模拟器中运行应用程序 11

1.5 Android 项目结构 🎥 12
    1.5.1 目录结构 12
    1.5.2 Android 应用程序结构分析 17

1.6 Android 应用程序设计示例 🎥 19

习题 1 21

第 2 章 Android 用户界面设计 ··· 22

- 2.1 用户界面设计和 View 类 ··· 22
- 2.2 Android 布局管理 ··· 22
  - 2.2.1 布局文件的规范与重要属性 ··· 23
  - 2.2.2 常见的布局方式 ··· 24
- 2.3 文本标签和按钮 ··· 32
  - 2.3.1 文本标签 ··· 32
  - 2.3.2 按钮及按钮处理事件 ··· 36
- 2.4 文本编辑框 ··· 38
- 2.5 进度条和选项按钮 ··· 42
  - 2.5.1 进度条 ··· 42
  - 2.5.2 选项按钮 ··· 44
- 2.6 图像显示类 ··· 51
- 2.7 消息提示类 ··· 56
- 2.8 列表组件类 ··· 59
- 习题 2 ··· 62

第 3 章 多个用户界面的程序设计 ··· 64

- 3.1 页面切换与传递参数值 ··· 64
  - 3.1.1 绑定机制组件 ··· 64
  - 3.1.2 Activity 页面切换 ··· 64
  - 3.1.3 在 Activity 页面之间传递数据 ··· 68
- 3.2 菜单设计 ··· 71
  - 3.2.1 选项菜单 ··· 72
  - 3.2.2 上下文菜单 ··· 74
- 3.3 对话框 ··· 76
  - 3.3.1 消息对话框 ··· 77
  - 3.3.2 其他几种常用对话框 ··· 81
- 3.4 Fragment ··· 85
  - 3.4.1 动态加载 Fragment 对象 ··· 85
  - 3.4.2 底部导航栏 ··· 88
- 习题 3 ··· 93

第 4 章 图形与多媒体处理 ··· 94

- 4.1 绘制几何图形 ··· 94
  - 4.1.1 几何图形绘制类 ··· 94
  - 4.1.2 几何图形绘制过程 ··· 95
  - 4.1.3 自定义组件 ··· 97

4.2 触摸屏事件处理 ·· 99
    4.2.1 简单的触摸屏事件 ·· 100
    4.2.2 手势识别 ·· 102
4.3 音频播放 ·· 108
    4.3.1 多媒体处理包 ·· 108
    4.3.2 多媒体处理播放器 ·· 108
    4.3.3 播放音频文件 ·· 109
4.4 视频播放 ·· 117
    4.4.1 应用媒体播放器播放视频 ·· 117
    4.4.2 应用视频视图播放视频 ·· 120
4.5 文本转换语音 ·· 124
4.6 动画技术 ·· 126
    4.6.1 动画组件类 ·· 126
    4.6.2 补间动画 ·· 127
    4.6.3 属性动画 ·· 133
习题 4 ·· 135

第 5 章 后台服务与系统服务 ·· 136

5.1 后台服务 ·· 136
5.2 信息广播机制 ·· 140
5.3 系统服务 ·· 147
    5.3.1 Android 的系统服务 ·· 147
    5.3.2 系统通知服务 ·· 147
习题 5 ·· 150

第 6 章 网络通信 ·· 151

6.1 Web 视图 ·· 151
    6.1.1 浏览器引擎 ·· 151
    6.1.2 Web 视图对象 ·· 151
    6.1.3 调用 JavaScript ·· 155
6.2 基于 TCP 的网络程序设计 ··· 160
    6.2.1 网络编程的基础知识 ·· 160
    6.2.2 利用套接字 Socket 设计客户端/服务器系统程序 ·········· 165
    6.2.3 应用 Callable 接口实现多线程 Socket 编程 ·················· 170
6.3 基于 HTTP 的网络程序设计 ·· 174
    6.3.1 建立 PHP 服务器网站 ·· 174
    6.3.2 应用 HttpURLConnection 访问 Web 服务器 ·················· 175
习题 6 ·· 184

第 7 章 应用 Volley 框架访问 Web 服务器 ··· 185

7.1 Volley 框架及其应用 ··· 185
7.1.1 Volley 包的下载与安装 ··· 185
7.1.2 JSON 数据格式简介 ··· 185
7.1.3 Volley 的工作原理和几个重要对象 ··· 190
7.1.4 Volley 的基本使用方法 ··· 191

7.2 应用 Volley 框架设计网络音乐播放器 ··· 196

7.3 访问远程数据库 ··· 201
7.3.1 把数据写入远程数据库 ··· 201
7.3.2 读取远程数据库数据 ··· 205

习题 7 ··· 209

第 8 章 数据存储 ··· 210

8.1 内部存储空间和外部存储空间 ··· 210

8.2 SQLite 数据库 ··· 212
8.2.1 SQLite 数据库简介 ··· 212
8.2.2 管理和操作 SQLite 数据库的对象 ··· 213
8.2.3 SQLite 数据库的操作命令 ··· 214

8.3 文件处理 ··· 225
8.3.1 输入/输出流 ··· 225
8.3.2 处理文件流 ··· 226

8.4 轻量级存储 SharedPreferences ··· 233

习题 8 ··· 235

第 9 章 OpenCV 应用实战——人脸美颜与人脸检测 ··· 237

9.1 OpenCV 图像处理 ··· 237
9.1.1 搭建 OpenCV Android 开发环境 ··· 237
9.1.2 Mat 对象和 Bitmap 对象 ··· 239
9.1.3 图像的模糊与锐化 ··· 242

9.2 人脸美颜 ··· 245

9.3 人脸检测 ··· 249

附表 微课视频二维码索引列表 ··· 253

第1章 Android系统及其开发过程

1.1 Android 系统概述

1.1.1 Android 系统及其特点

2007 年 11 月 5 日，Google 公司推出了基于 Linux 操作系统的智能手机平台 Android 系统。Android 系统由操作系统、中间件、用户界面程序和应用软件等组成。2013 年 5 月 16 日，Google 公司推出新的 Android 开发环境——Android Studio。Android 的出现绝非偶然，它是由传统的移动电话系统开发模式演变而来的一种符合时代潮流的新型移动开发模式的产物。

Android 系统诞生在开放时代的背景下，其全开放的智能移动平台、多硬件平台的支持、使用众多标准化的技术、核心技术完整、完善的 SDK 和文档、完善的辅助开发工具等特点与智能手机发展方向紧密相连，它将代表并引领着新时代的技术潮流。

Android 系统具有开放性、平等性、方便性及硬件丰富性等特点。下面对这些特点进行简单介绍。

1．系统开放性

Android 系统是一款真正开放的系统。Android 系统从底层的操作系统直到最上层的应用程序都是开放的，程序开发人员和爱好者都可以很方便地从网络上获取源代码，可以对它们进行分析和移植。

2．应用程序平等性

在 Android 系统开发平台上，Android 系统自带的程序与程序开发人员自己开发的应用程序都是平等的，程序开发人员可以开发个人喜爱的应用程序来替代系统的程序，构建个性化的 Android 系统。

3．开发方便性

在 Android 系统开发平台上开发应用程序是非常方便的，Android 系统为开发人员提供了大量的实用组件库和方便的工具，开发人员只需编写几行代码就可以将功能强大的组件添加到自己的程序中。

4. 硬件丰富性

由于 Android 系统的开放性，众多的硬件制造商纷纷开发出各种各样的可以与 Android 系统兼容的产品，进一步丰富了 Android 系统的应用。

■ 1.1.2 Android 的系统架构

Android 的系统架构和其操作系统一样，采用了分层的架构。Android 系统分为 4 层，从顶层到底层分别是应用程序层、应用程序框架层、系统运行库层和 Linux 核心层，如图 1.1 所示。

1. 应用程序

Android 系统自带了一套核心应用程序，包括电话拨号程序、短信程序、日历、音乐播放器、浏览器、联系人管理程序等，如图 1.2 所示。所有的应用程序都是用 Java 语言编写的，开发人员自己开发的应用程序就位于该层。该层的程序是完全平等的，开发人员可以任意用自己开发的应用程序替换 Android 系统自带的程序。

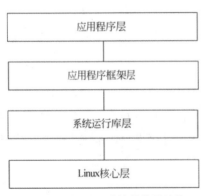

图 1.1 Android 系统的体系结构

图 1.2 Android 系统自带的应用程序

2．应用程序框架

Android 系统通过应用程序框架为开发人员创建自己的应用程序提供了一个开放的开发平台，程序开发人员可以在这个应用程序框架平台上设计自己的应用程序。本书所讲的程序设计都是基于这个应用程序框架来完成设计的。

Android 系统的应用程序框架主要包含以下 9 个部分。

（1）活动页面管理（Activity Manager）：用于管理程序的生命周期。

（2）窗口管理（Window Manager）：用于管理应用程序窗口。

（3）内容供应（Content Provider）：提供数据共享，使一个应用程序可以访问另一个应用程序的数据。

（4）视图系统（View System）：用于构建应用程序的可视化组件。

（5）包管理（Package Manager）：用于管理项目程序。

（6）电话管理（Telephone Manager）：移动设备的基本功能统一由电话管理器管理。

（7）资源管理（Resource Manager）：为应用程序提供所需的文字、声音、图片、视频或布局文件等资源。

（8）位置管理（Location Manager）：用于提供位置服务。

（9）通知管理（Notification Manager）：在手机顶部状态栏发布消息提示。

3．系统运行库

（1）程序库。

Android 包含一些 C/C++ 程序库，这些库能被 Android 系统中不同的组件使用。它们通过 Android 应用程序框架为开发者提供服务。

（2）Android 运行时库。

Android 包含一个核心库，该核心库提供了 Java 编程语言核心库的大多数功能。Android 系统的 Dalvik 虚拟机也包含在 Android 运行时库中。

4．Linux 核心

Android 的核心系统服务依赖于 Linux 核心，其安全性、内存管理、进程管理、网络协议栈和驱动模型等基本依赖于 Linux。

1.1.3　Android 开发分类

对于开发者而言，Android 开发分为以下两大类。

1．系统移植开发

移植开发是为了使 Android 系统能在手持式移动设备上运行，在具体的硬件系统上构建 Android 软件系统。这种类型的开发在 Android 底层进行，需要移植开发 Linux 中相关的设备驱动程序及 Android 本地框架中的硬件抽象层，也就是需要将设备驱动与 Android 系统联系起来。Android 系统对硬件抽象层都有标准的接口定义，在移植时只需实现这些接口即可。

2. Android 应用程序开发

应用程序开发既可以基于硬件设备（用于测试的实体手机），也可以基于 Android 模拟器。应用开发处于 Android 系统的顶层，使用 Android 系统提供的 Java 框架（API）进行开发设计工作，是大多数开发者从事的开发工作。本书所介绍的 Android 应用程序设计都是在这个层次上进行的。

1.2 搭建 Android Studio 开发环境

1.2.1 安装 Android Studio 前的必要准备

1. Android 系统开发的操作平台

Android 系统开发的软件环境目前有两种，一种是 Eclipse + ADT（Android Development Tools 插件）系统，另一种是 Android Studio 系统。在这里主要介绍 Android Studio 系统。

Android Studio 是一个全新的基于 IntelliJ IDEA 的 Android 开发环境（IntelliJ IDEA 是一种由 Java 语言开发的集成开发环境，是被业界公认为优秀的 Java 开发工具），Android Studio 提供了集成的 Android 开发工具用于应用程序的开发和调试。

在安装 Android Studio 之前，需要安装 Java JDK 的环境。

2. 下载最新版本的 Android Studio 软件

读者可以访问 Android Studio 官方网站 https://developer.android.google.cn/studio 下载最新的系统软件，如图 1.3 所示。

图 1.3 Android Studio 官方下载首页

进入下载页面以后，下载对应操作系统所支持的版本（以 Android Studio 2020.3.1 版本为例），如表 1-1 所示。

表 1-1　不同平台下载 Android Studio 安装包版本

安 装 平 台	系统安装包	所 需 空 间
Windows（64 位）	android-studio-2020.3.1.24-windows.exe 包含 Android SDK（推荐）	912 MB
	android-studio-2020.3.1.24-windows.zip	922 MB
macOS X	android-studio-2020.3.1.24-mac.dmg	947 MB
Linux	android-studio-2020.3.1.24-linux.tar.gz	935 MB

■ 1.2.2　安装 Android Studio

1. 按照安装向导完成 Android Studio 系统的安装

运行安装文件 android-studio-2020.3.1.24-windows.exe，按照安装向导完成系统的安装，如图 1.4 所示。

图 1.4　Android Studio 系统的安装

系统安装完成的提示界面如图 1.5 所示。

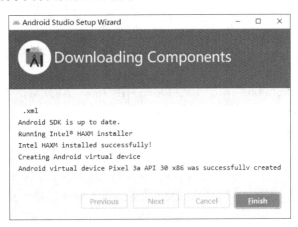

图 1.5　Android Studio 系统安装完成的提示界面

2. 设置 Android SDK 存放位置

安装完成后，第一次运行 Android Studio 系统需要设置 Android SDK，找到 SDK 的存放位置，如图 1.6 所示。Android SDK 的存放位置也可以通过 Android Studio 应用程序的 Settings 命令设置。

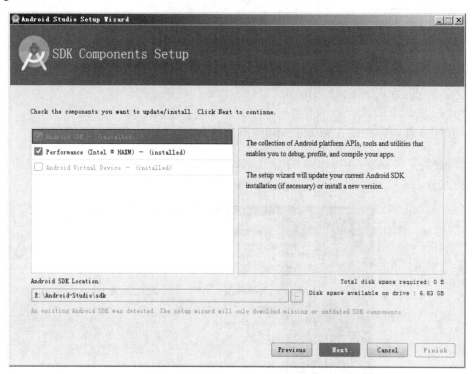

图 1.6　设置 Android SDK 的位置

3. 创建 Android 虚拟设备

Android 应用程序既可以在实体手机上执行，也可以创建一个 Android 虚拟设备（Android Virtual Device，AVD）来测试。每个 AVD 模拟一套虚拟环境来运行 Android 操作系统平台，这个平台有自己的内核、系统图像、外观显示、用户数据区和仿真的 SD 卡等。

下面介绍如何创建一个 AVD。

Android Studio 集成开发环境提供了 Android Virtual Device Manager 功能，可以采用它来创建和调用 AVD。

（1）选择 Android Studio 菜单栏中的 Tools→Android→AVD Manager 命令，用户在弹出的 Android Virtual Device Manager 对话框中可以看到已创建的 AVD，如图 1.7 所示。单击下方的 Create Virtual Device 按钮创建一个新的 AVD。

（2）运行 AVD 模拟器。在 Android Virtual Device Manager 对话框中选择已经建立的 AVD，单击 Actions 栏中的 ▶ 按钮可以启动 AVD 模拟器。启动 AVD 模拟器的时间很长，建议打开后不要关闭，可以在该模拟器上测试 Android 应用程序。启动的 AVD 模拟器如图 1.8 所示。

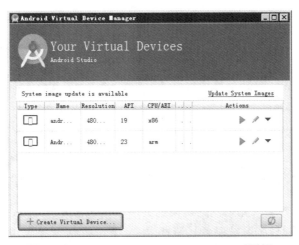

图 1.7 Android Virtual Device Manager 对话框

图 1.8 Android 的 AVD 模拟器

1.3 Android API 和在线帮助文档

1. Android API

Android 为用户安装了它所提供的标准类库。所谓标准类库，就是把程序设计所需要的常用的方法和接口分类封装成包，Android 所提供的标准类库就是 Android API。

Android 包中封装了程序设计所需要的主要应用类，本书中所用到的包如下。

- Android.app：封装了高层的程序模型，提供基本的运行环境。
- Android.content：封装了各种对设备上的数据进行访问和发布的类。
- Android.database：通过内容提供者浏览和操作数据库。
- Android.graphics：底层的图形库，包含画布、颜色过滤、点、矩形，可以将它们直接绘制到屏幕上。
- Android.location：封装了定位和相关服务的类。
- Android.media：封装了一些类管理多种音频、视频的媒体接口。
- Android.net：封装了帮助网络访问的类，超过通常的 java.net.*接口。
- Android.os：封装了系统服务、消息传输、IPC 机制。
- Android.opengl：封装了 opengl 的工具、3D 加速。
- Android.provider：封装了类访问 Android 的内容提供者。
- Android.telephony：封装了与拨打电话相关的 API 交互。
- Android.view：封装了基础的用户界面接口框架。
- Android.util：涉及工具性的方法，例如时间、日期的操作。
- Android.webkit：默认浏览器操作接口。
- Android.widget：封装了各种 UI 元素（大部分是可见的），在应用程序的屏幕中使用。

2. Android API 帮助文档

Android Studio 提供了离线的 Android API 文档，这是进行程序设计的好工具，希望大家都能用好这个工具。

运行 Android Studio 安装目录下的 index.html 文件，运行结果如图 1.9 所示。

图 1.9 Android API 在线帮助文档

1.4 Android 应用程序的开发过程

1.4.1 开发 Android 应用程序的一般过程

开发 Android 应用程序的一般过程如图 1.10 所示。

图 1.10　Android 应用程序的开发过程

1.4.2 生成 Android 应用程序框架

1. 创建一个新的 Android 项目

启动 Android Studio，在弹出的对话框中选择 Create New Project 选项，创建一个新的 Android 项目，如图 1.11 所示。

图 1.11　新建一个 Android 项目

2. 填写应用程序的参数

在"Android 新应用程序"对话框中按步骤选择 Android Studio 的应用程序 Empty Activity 模板，并设置应用程序名称、包名、项目保存路径等参数，如图 1.12 和图 1.13 所示。

图 1.12　选择 Empty Activity 模板

图 1.13　设置 Android 项目参数

最后系统自动生成一个 Android 应用项目框架，如图 1.14 所示。

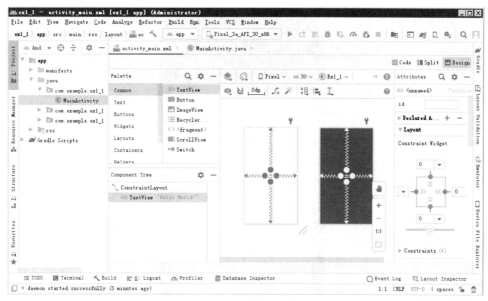

图 1.14　系统自动生成的 HelloAndroid 应用项目框架

1.4.3　编写代码生成 MainActivity.java

创建 HelloAndroid 项目后，打开主程序文件 MainActivity.java，可以看到系统自动生成的代码如下：

```
1    package com.example.helloandroid;
2    import androidx.appcompat.app.AppCompatActivity;
3    import android.os.Bundle;

4    public class MainActivity extends AppCompatActivity
5    {
6      @Override
7      public void onCreate(Bundle savedInstanceState)
8      {
9        super.onCreate(savedInstanceState);
10       setContentView(R.layout.activity_main);
11     }
12   }
```

显示 activity_main.xml 定义的用户界面

在 Android 系统中，应用程序的入口程序（主程序）都是活动程序界面 Activity 类的子类。在上述代码中，最重要的是第 10 行，用于显示用户界面。

1.4.4　在模拟器中运行应用程序

在工具栏中单击 Run App 按钮，运行 AVD 模拟器，可以看到应用程序的运行结果

（首次运行程序可能耗时较长），如图 1.15 所示。

图 1.15 在 AVD 模拟器上显示程序运行结果

1.5 Android 项目结构

1.5.1 目录结构

打开 HelloAndroid 项目，在项目资源管理器中可以看到应用项目的目录和文件结构，如图 1.16 所示。

实际上，图 1.16 所示的目录结构内容是最基本的，程序员还可以在此基础上添加需要的内容。下面对 app 模块下的文件目录结构的基本内容进行介绍。

（1）manifests：其下的 AndroidManifest.xml 为项目的配置信息文件。

（2）java：主要是源代码和测试代码。

（3）res：主要是资源目录，存储所有的项目资源。

（4）values：存储 app 引用的一些值。

- colors.xml：存储了一些 color 的样式。
- dimens.xml：存储了一些公用的 dip 值。
- strings.xml：存储了引用的 string 值。
- styles.xml：存储了 app 需要用到的一些样式。

（5）Gradle Scripts：build.gradle 为项目的 gradle 配置文件。

1. 源代码目录 java

java 目录存放 Android 应用程序的 Java 源代码文件。在系统自动生成的项目结构中，有一个在创建项目时输入 Create Activity 名称的 Java 文件 MainActivity.java，如图 1.17 所示。

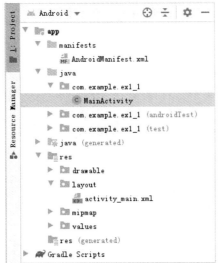

图 1.16 HelloAndroid 项目的目录和文件结构　　图 1.17 src 目录下的 MainActivity.java 源代码

2. 资源目录 res 和资源类型

Android 系统的资源为应用项目所需要的声音、图片、视频、用户界面文档等，其资源文件存放于项目的 res 目录下。资源的目录结构及类型如表 1-2 所示。

表 1-2 Android 系统资源的目录结构及类型

目 录 结 构	资 源 类 型
res/drawable	图片（bmp、png、gif、jpg 等）
res/layout	XML 界面布局文件
res/mipmap	存储系统的图片资源
res/values	存放字符串、颜色、尺寸、数组、主题、类型等资源
res/raw	可以存放任意类型的文件，一般存放比较大的音频、视频、图片或文档，会在 R 类中生成资源 id，封装在 apk 中
assets	可以存放任意类型，不会被编译，与 raw 相比，不会在 R 类中生成资源 id

（1）目录 mipmap 存储系统的图片资源，*dpi 表示存储不同分辨率的图片，分别为分辨率大小不同的图标资源，以便相同的应用程序在分辨率大小不同的显示窗体上都可以顺利显示。系统开始运行时会检测显示窗体的分辨率大小，自动选择与显示窗体分辨率大小匹配的目录，获取大小匹配的图标，如表 1-3 所示。

（2）在 layout 子目录中存放用户界面布局文件。该子目录中有一个系统自动生成的 activity_main.xml 文件，它既可以按可视化的图形设计界面显示，也可以按代码设计界面显示，如图 1.18 所示。

表 1-3　4 种分辨率大小不同的图标

图标文件类型	像 素 密 度	图标分辨率
mdpi	中等密度	320 像素×480 像素
hdpi	高密度	480 像素×800 像素
xhdpi	超高密度	960 像素×720 像素
xxhdpi	超超高密度	1280 像素×720 像素

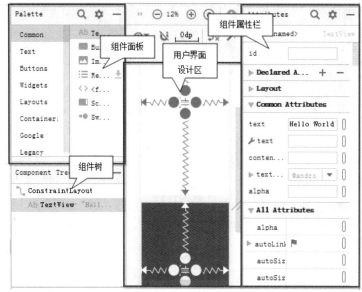

（a）图形设计界面

（b）代码设计界面

图 1.18　用户界面布局文件 **activity_main.xml**

activity_main.xml 布局文件的代码如下：

```
1   <?xml version="1.0" encoding="utf-8"?>
2   <androidx.constraintlayout.widget.ConstraintLayout
    xmlns:android="http://schemas.android.com/apk/res/android"
3       xmlns:app="http://schemas.android.com/apk/res-auto"
4       xmlns:tools="http://schemas.android.com/tools"
5       android:layout_width="match_parent"
6       android:layout_height="match_parent"
7       tools:context=".MainActivity">
8
9       <TextView
10          android:layout_width="wrap_content"
11          android:layout_height="wrap_content"
12          android:text="Hello World!"
13          android:textSize="32sp"
14          app:layout_constraintBottom_toBottomOf="parent"
15          app:layout_constraintLeft_toLeftOf="parent"
16          app:layout_constraintRight_toRightOf="parent"
17          app:layout_constraintTop_toTopOf="parent" />
18
19  </androidx.constraintlayout.widget.ConstraintLayout>
```

布局参数解析：

- "<ConstraintLayout>"：约束布局配置，在约束布局中，所有组件都需要通过建立约束来确定摆放位置。
- "android:layout_width"：定义当前视图在屏幕上所占的宽度，match_parent 即填充整个屏幕宽度。
- "android:layout_height"：定义当前视图在屏幕上所占的高度。
- "wrap_content"：自适应大小，以便显示其全部文字内容。

在应用程序中需要使用用户界面的组件时，需要通过 R.java 文件中的 R 类来调用。

（3）values 子目录存放参数描述文件资源。这些参数描述文件也都是 XML 文件，如字符串（strings.xml）、颜色（colors.xml）、数组（arrays.xml）等。这些参数也需要通过 R.java 文件中的 R 类来调用。

3．资源索引文件 R.txt

在 app\build\intermediates\runtime_symbol_list\debug\目录下存放由系统自动产生的一个 R.txt 文件，该文件将 res 目录中的资源与编号（id）进行映射，从而可以方便地对资源进行引用，如图 1.19 所示。正如该文件头部注释的说明，该文件是自动生成的，不允许用户修改。

在程序中引用资源需要使用 R 类，其引用形式如下：

R.资源类型.资源名称

例如：

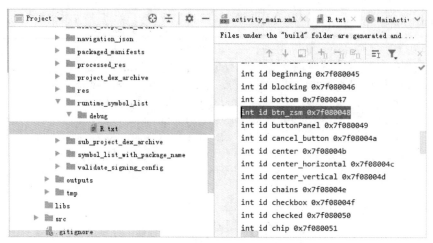

图 1.19 资源索引文件 R.txt 源代码

（1）在 Activity 中显示布局视图：

setContentView(R.layout.main);

（2）程序要获得用户界面布局文件中的按钮实例 Button1：

mButtn = (Button)findViewById(R.id.Button1);

（3）程序要获得用户界面布局文件中的文本组件实例 TextView1：

mEditText = (EditText)findViewById(R.id.EditText1);

在编写和调试程序过程中，有时由于操作失误会造成 MainActivity.java 程序中找不到 R 文件或显示 R 文件错误。如果出现这种情况，可以按下列步骤解决。

（1）检查资源文件是否存在错误，包括 layout 文件以及图片资源等文件，如有错误，及时更正。

（2）执行 Android Studio 菜单栏中的 Build→Clean Project 命令，如图 1.20 所示，经项目清理后，MainActivity.java 程序中找不到 R 文件或显示 R 文件错误的问题就能解决。

图 1.20 清理项目

4. 项目配置文件 AndroidManifest.xml

AndroidManifest.xml 是每个应用程序都需要的系统配置文件，它位于应用程序根目录下。AndroidManifest.xml 文件的代码如下：

```
1    <?xml version="1.0" encoding="utf-8"?>
2    <manifest xmlns:android="http://schemas.android.com/apk/res/android"
3        package="com.HelloAndroid"
4        android:versionCode="1"
5        android:versionName="1.0">
6        <uses-sdk android:minSdkVersion="14" />
```

```
7      <application
8          android:icon="@drawable/ic_launcher"
9          android:label="@string/app_name">
10         <activity
11           android:label="@string/app_name"
12           android:name=".MainAndroidActivity">
13           <intent-filter >
14             <action android:name="android.intent.action.MAIN" />
15             <category android:name="android.intent.category.LAUNCHER" />
16           </intent-filter>
17         </activity>
18       </application>
19  </manifest>
```

AndroidManifest.xml 文件代码元素的说明如表 1-4 所示。

表 1-4　AndroidManifest.xml 文件代码说明

代码元素	说　　明
manifest	XML 文件的根结点，包含了 package 中所有的内容
xmlns:android	命名空间的声明。 xmlns:android="http://schemas.android.com/apk/res/android"使得 Android 中各种标准属性能在文件中使用
package	声明应用程序包
uses-sdk	声明应用程序所使用的 Android SDK 版本
application	application 级别组件的根结点，声明一些全局或默认的属性，如标签、图标、必要的权限等
android:icon	应用程序图标
android:label	应用程序名称
activity	Activity 是一个应用程序与用户交互的图形界面。每一个 Activity 必须有一个<activity>标记对应，如果一个 Activity 没有对应的标记将无法运行
android:name	应用程序默认启动的活动程序的 Activity 界面
intent-filter	声明一组组件支持的 Intent 值。在 Android 中,组件之间可以相互调用、协调工作,Intent 提供组件之间通讯所需要的相关信息
action	声明目标组件执行的 Intent 动作。Android 定义了一系列标准动作，如 MAIN_ACTION、VIEW_ACTION、EDIT_ACTION 等。与此 Intent 匹配的 Activity 将会被当作进入应用的入口
category	指定目标组件支持的 Intent 类别

1.5.2　Android 应用程序结构分析

1．逻辑控制层与表现层

从上面的 Android 应用程序可以看到，一个 Android 应用程序通常由 Activity 类程序（Java 程序）和用户界面布局 XML 文档组成。

在 Android 应用程序中，逻辑控制层与表现层是分开设计的。逻辑控制层由 Java 应用程序实现，表现层由 XML 文档描述，如图 1.21 所示。

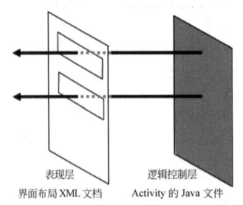

图 1.21　Android 应用程序的逻辑控制层与表现层

2. Android 程序的组成结构

Android 程序与 Java 程序的结构是相同的，打开 src 目录下的 HelloAndroid.java 文件，其代码如下：

```
1   package com.example.HelloAndroid;          ← 包声明语句

2   import androidx.app.AppCompatActivity;
3   import android.os.Bundle;                   ← 导入包

                                                类标志
4   public class MainAndroid extends AppCompatActivity    ← 类声明语句
5   {                                                类名
6       public void onCreate(Bundle savedInstanceState)   ← 重写 onCreate()方法
7       {
8           super.onCreate(savedInstanceState);   ← 调用父类 Activity 的 onCreate()方法
9           setContentView(R.layout. activity_main);
10      }                                         ← 在屏幕上显示内容的方法
11  }
```

其中：

（1）第 1 行是包声明语句，包名字是在建立应用程序的时候指定。在这里设定为 package com.example.HelloAndroid。该行的作用是指出这个文档所在的名称空间，package（包）是其关键字。使用名称空间的原因是程序一旦扩展到某个大小，程序中的变量名称、方法名称、类名等难免重复，这时就可以通过定义名称空间将定义的名称区隔，以避免相互冲突的情形发生。

（2）第 2 行和第 3 行是导入包的声明语句。这两条语句的作用是告诉系统编译器，编译程序时要导入 androidx.app.AppCompatActivity 和 android.os.Bundle 两个包。import（导入）是其关键字。在 Java 语言中，使用任何 API 都要事先导入相对应的包。

（3）第 4～11 行是类的定义，这是应用程序的主体部分。Android 应用程序是由类组成的，类的一般结构如下：

```
public class MainAndroid extends AppCompatActivity  //类声明
```

```
{
    … ;   // 类体
}
```

class 是类的关键字，HelloAndroid 是类名。在 public class MainAndroid 后面添加 extends AppCompatActivity，则表示 MainAndroid 类继承 AppCompatActivity 类。这时称 AppCompatActivity 类是 HelloAndroid 类的父类，或称 HelloAndroid 类是 AppCompatActivity 类的子类。extends 是表示继承关系的关键字。在面向对象的程序中，子类会继承父类所有方法和属性。也就是说，对于在父类中所定义的全部方法和属性，子类可以直接拿来使用。由于 AppCompatActivity 类是一个具有屏幕显示功能的活动界面程序，因此其子类 MainAndroid 也具有屏幕显示功能。

class 语句后面跟着的一对大括号"{ }"表示复合语句，是该类的主体部分，称为类体。在类体中定义类的方法和变量。

（4）第 6～10 行是在 MainAndroid 类的类体中定义一个方法。

1.6 Android 应用程序设计示例

视频讲解

【例 1-1】 在 AVD 模拟器中显示"我对学习 Android 很感兴趣！"。

（1）在 Android Studio 中新建一个 Android 项目，其项目名称（Application Name）为 ex1_1。

（2）在系统自动生成的应用程序框架中打开修改资源目录（res\layout）中的界面布局文件 activity_main.xml，在设计界面中选中文本标签组件 TextView（组件面板中的 TextView 项），在属性栏中找到 text 属性，将其修改为"我对学习 Android 很感兴趣"，如图 1.22（a）所示。

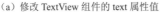

(a) 修改 TextView 组件的 text 属性值　　　　(b) 在模拟器中运行的结果

图 1.22　在模拟器中显示"我对学习 Android 很感兴趣"

（3）保存程序后，单击工具栏中的 Run App 按钮 运行 AVD 模拟器，在模拟器中的程序运行结果如图 1.22（b）所示。

【例 1-2】 设计一个显示资源目录中图片文件的程序。

（1）在 Android Studio 中新建一个 Android 项目，其项目名称（Application Name）为 ex1_2。

（2）把事先准备的图片文件 flower.png 复制到资源目录 res\drawable 中，如图 1.23（a）所示。

（3）打开源代码目录 src 中的 MainActivity.java 文件，编写代码如下。

```
1   package com.example.ex1_2;
2   import androidx.appcompat.app.AppCompatActivity;
3   import android.os.Bundle;
4   import android.widget.ImageView;          ← 增加导入 ImageView 类的语句
5   public class MainActivity extends AppCompatActivity {
6     @Override
7     public void onCreate(Bundle savedInstanceState) {
8       super.onCreate(savedInstanceState);
9       // setContentView(R.layout.activity_main);   ← 注释该语句
10      ImageView img = new ImageView(this);   //创建 ImageView 对象并实例化
11      img.setImageResource(R.drawable.flower);//ImageView 对象设置引用图片资源
12      setContentView(img);  ← 把 ImageView 对象显示到屏幕上
13    }
14  }
```

（4）保存程序。选择菜单"运行"的"运行配置"命令，运行项目。在模拟器中运行的结果如图 1.23（b）所示。

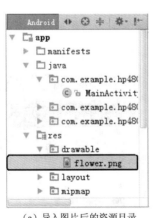
（a）导入图片后的资源目录

（b）程序在模拟器中运行的结果

图 1.23　在模拟器中显示图片

习题 1

1. Android 系统是基于什么操作系统的应用系统？
2. 试述建立 Android 系统开发环境的过程和步骤。
3. 如何编写和运行 Android 系统应用程序？
4. 编写 Android 应用程序，在模拟器中显示"我对 Android 很痴迷！"。
5. 编写 Android 应用程序，在模拟器中显示一个图片文件。

第2章 Android用户界面设计

2.1 用户界面设计和 View 类

在 Android 应用中，用户界面（User Interface，UI）是人与手机之间数据传递、交互信息的重要对话接口，是 Android 系统的重要组成部分。

一个 Android 应用的界面由 View 类和 ViewGroup 对象构建，而 ViewGroup 对象都是 View 类的子类。View 类是所有可视化组件的共同父类，几乎所有的用户界面组件都是继承 View 类而实现的，如 TextView、Button、EditText 等。

对 View 类及其子类的属性进行设置，既可以在布局文件 XML 中设置，也可以通过成员方法在 Java 代码文件中动态设置。View 类常用的属性和方法如表 2-1 所示。

表 2-1 View 类的常用属性和方法

属　　性	对　应　方　法	说　　明
android:background	setBackgroundColor(int color)	设置背景颜色
android:id	setId(int)	为组件设置可通过 findViewById 方法获取的标识符
android:alpha	setAlpha(float)	设置透明度，取值范围为 0～1
android:view	findViewById(int id)	与 id 所对应的组件建立关联
android:visibility	setVisibility(int)	设置组件的可见性
android:clickable	setClickable(boolean)	设置组件是否响应单击事件

视频讲解

2.2 Android 布局管理

Android 系统按照 MVC（Model-View-Controller）设计模式，将应用程序的界面设计与功能控制设计分离，从而可以单独地修改用户界面，而不需要去修改程序代码。应用程序的用户界面通过 XML 定义组件布局来实现。

Android 系统的布局管理指的是在 XML 布局文件中设置组件的大小、间距、排列及对齐方式等。Android 系统中常见的布局方式有 5 种，分别是 ConstraintLayout、LinearLayout、

FrameLayout、TableLayout 和 GridLayout。组件面板中的布局组件如图 2.1 所示。

图 2.1　布局组件

■ 2.2.1　布局文件的规范与重要属性

1．布局文件的规范

Android 系统应用程序的 XML 布局文件有以下规范。
（1）布局文件作为应用项目的资源存放在 res/layout 目录下，其扩展名为.xml。
（2）布局文件的根节点通常是一个布局方式，在根节点内可以添加组件作为节点。
（3）布局文件的根节点必须包含一个命名空间：

```
xmlns:android="http://schemas.android.com/apk/res/android"
```

（4）如果要在实现控制功能的 Java 程序中控制界面中的组件，则必须为界面文件中的组件定义一个 id，其定义格式为：

```
android:id="@+id/<组件 ID>"
```

2．布局文件的重要属性值

一个界面布局中会有很多元素，这些元素的大小和位置由其属性决定。下面简要介绍布局文件中的几个重要属性值。
（1）设置组件大小的属性值。
- wrap_content：根据组件内容的大小来决定组件的大小。
- match_parent：使组件填充所在容器的所有空间。

（2）设置组件大小的单位。
- px（pixels）像素：即屏幕上的发光点。
- dp（或 dip，全称为 device independent pixels）设备独立像素：一种支持多分辨率设备的抽象单位，和硬件相关。
- sp（scaled pixels）比例像素：设置字体大小。

（3）设置组件的对齐方式。

在布局文件中，由 android:gravity 属性控制组件的对齐方式，其属性值有上（top）、下（bottom）、左（left）、右（right）、水平方向居中（center_horizontal）、垂直方向居中（center_vertical）等。

2.2.2 常见的布局方式

1. 约束布局

约束布局（ConstraintLayout）是采用相对其他组件的位置的布局方式，这是 Android Studio 系统默认的布局方式。

在约束布局中，通过指定 id 关联其他组件，与之右对齐、上下对齐或屏幕中央对齐等方式来排列组件。

在 XML 布局文件中，由根元素 ConstraintLayout 来标识约束布局。约束布局的属性如下。

  layout_constraintTop_toTopOf：该组件的上边对齐另一个组件的上边。
  layout_constraintTop_toBottomOf：该组件的上边对齐另一个组件的底边。
  layout_constraintTop_toLeftOf：该组件的上边对齐另一个组件的左边。
  layout_constraintTop_toRightOf：该组件的上边对齐另一个组件的右边。
  layout_constraintBottom_toTopOf：该组件的下边对齐另一个组件的上边。
  layout_constraintBottom_toBottomOf：该组件的底边对齐另一个组件的底边。
  layout_constraintBottom_toLeftOf：该组件的底边对齐另一个组件的左边。
  layout_constraintBottom_toRightOf：该组件的底边对齐另一个组件的右边。
  layout_constraintLeft_toTopOf：该组件的左边对齐另一个组件的上边。
  layout_constraintLeft_toBottomOf：该组件的左边对齐另一个组件的底边。
  layout_constraintLeft_toLeftOf：该组件的左边对齐另一个组件的左边。
  layout_constraintLeft_toRightOf：该组件的左边对齐另一个组件的右边。
  layout_constraintRight_toTopOf：该组件的右边对齐另一个组件的上边。
  layout_constraintRight_toBottomOf：该组件的右边对齐另一个组件的底边。
  layout_constraintRight_toLeftOf：该组件的右边对齐另一个组件的左边。
  layout_constraintRight_toRightOf：该组件的右边对齐另一个组件的右边。

把组件拖到界面中，它会在每个角上显示一个带有方形调整大小控键的边框，并在每边显示圆形约束控键，可以通过拖曳实现调整组件大小及添加约束，如图 2.2 所示。

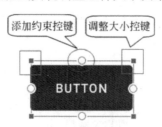

图 2.2 约束布局的调整控键

- 调整大小控键：可以拖动方形调整大小的句柄来调整元素的大小。
- 添加约束控键：单击约束句柄，在元素的每一侧显示为圆形，然后拖动到另一个组件的约束句柄或父边界以创建约束。约束由 Z 字形线表示。

【例 2-1】 约束布局应用示例。

创建名称为 ex2_1 的新项目，包名为 com.example.ex2_1。打开系统自动生成的项目框架，在界面布局文件 activity_main.xml 中加入两个按钮组件 Button，设置约束如图 2.3 所示。

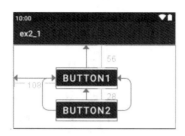

图 2.3　约束布局示例

界面布局程序 activity_main.xml 的代码如下：

```
1    <?xml version="1.0" encoding="utf-8"?>
2    <androidx.constraintlayout.widget.ConstraintLayout
3        xmlns:android="http://schemas.android.com/apk/res/android"
4        xmlns:app="http://schemas.android.com/apk/res-auto"
5        xmlns:tools="http://schemas.android.com/tools"
6        android:layout_width="match_parent"
7        android:layout_height="match_parent"
8        tools:context=".MainActivity"
9        tools:layout_editor_absoluteY="81dp">
```

```
10       <Button
11           android:id="@+id/button1"
12           android:layout_width="wrap_content"
13           android:layout_height="wrap_content"
14           android:layout_marginStart="108dp"
15           android:layout_marginLeft="108dp"
16           android:layout_marginTop="56dp"
17           android:text="Button1"
18           android:textSize="24sp"
19           app:layout_constraintStart_toStartOf="parent"
20           app:layout_constraintTop_toTopOf="parent" />

21       <Button
22           android:id="@+id/button2"
23           android:layout_width="wrap_content"
24           android:layout_height="wrap_content"
25           android:layout_marginTop="28dp"
26           android:text="Button2"
27           android:textSize="24sp"
28           app:layout_constraintEnd_toEndOf="@+id/button1"
29           app:layout_constraintHorizontal_bias="0.0"
30           app:layout_constraintStart_toStartOf="@+id/button1"
```

```
31          app:layout_constraintTop_toBottomOf="@+id/button1" />
32  </androidx.constraintlayout.widget.ConstraintLayout>
```

在上述代码中，灰色底纹的代码为系统自动生成，其他代码为拖进两个按钮并建立约束后生成的。

2. 线性布局

线性布局（LinearLayout）是 Android 系统中常用的布局方式之一，它将组件按照水平或垂直方向排列。在 XML 布局文件中，由根元素 LinearLayout 来标识线性布局。

在布局文件中，由 android:orientation 属性来控制排列方向，其属性值有水平（horizontal）和垂直（vertical）两种。

- 设置线性布局为水平方向：

```
android:orientation="horizontal"
```

- 设置线性布局为垂直方向：

```
android:orientation="vertical"
```

【例 2-2】 线性布局应用示例。

创建名称为 ex2_2 的新项目，包名为 com.example.ex2_2。生成项目框架后，修改布局文件 activity_main.xml 如下：

```
1   <?xml version="1.0" encoding="utf-8"?>
2   <androidx.constraintlayout.widget.ConstraintLayout
3       xmlns:android="http://schemas.android.com/apk/res/android"
4       xmlns:app="http://schemas.android.com/apk/res-auto"
5       xmlns:tools="http://schemas.android.com/tools"
6       android:layout_width="match_parent"
7       android:layout_height="match_parent"
8       tools:context=".MainActivity">
9       <LinearLayout
10          android:id="@+id/linearLayout"
11          android:layout_width="0dp"
12          android:layout_height="0dp"
13          android:orientation="vertical"
14          app:layout_constraintBottom_toBottomOf="parent"
15          app:layout_constraintEnd_toEndOf="parent"
16          app:layout_constraintStart_toStartOf="parent"
17          app:layout_constraintTop_toTopOf="parent">
18          <Button
19              android:id="@+id/button"
20              android:layout_width="150dp"
21              android:layout_height="wrap_content"
22              android:text="按钮 1" />
23          <Button
```

```
24          android:id="@+id/button2"
25          android:layout_width="150dp"
26          android:layout_height="wrap_content"
27          android:text="按钮2" />
28      <Button
29          android:id="@+id/button3"
30          android:layout_width="150dp"
31          android:layout_height="wrap_content"
32          android:text="按钮3" />
33  </LinearLayout>
34 </androidx.constraintlayout.widget.ConstraintLayout>
```

在上述代码中,灰色底纹的代码为系统自动生成,其他为拖进设计界面中组件并建立约束后由系统生成的。程序运行结果如图 2.4(a)所示。如果将代码中第 13 行的 android:orientation="vertical"(垂直方向的线性布局)更改为 android:orientation="horizontal"(水平方向的线性布局),则运行结果如图 2.4(b)所示。

 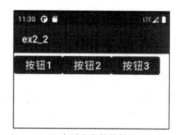

(a)垂直方向的线性布局　　　　　　(b)水平方向的线性布局

图 2.4　线性布局示例

3. 帧布局

帧布局(FrameLayout)是将组件放置到左上角位置,当添加多个组件时,后面的组件将遮盖之前的组件。

【例 2-3】　帧布局应用示例。

创建名称为 ex2_3 的新项目,包名为 com.example.ex2_3。生成项目框架后,将事先准备的图像文件 img.png 复制到 res\drawable 目录下。

界面布局文件 activity_main.xml 代码如下:

```
1  <?xml version="1.0" encoding="utf-8"?>
2  <androidx.constraintlayout.widget.ConstraintLayout
3      xmlns:android="http://schemas.android.com/apk/res/android"
4      xmlns:app="http://schemas.android.com/apk/res-auto"
5      xmlns:tools="http://schemas.android.com/tools"
6      android:layout_width="match_parent"
7      android:layout_height="match_parent"
8      tools:context=".MainActivity">

9      <FrameLayout
```

```
10        android:id="@+id/frameLayout "
11        android:layout_width="0dp"
12        android:layout_height="0dp"
13        app:layout_constraintBottom_toBottomOf="parent"
14        app:layout_constraintEnd_toEndOf="parent"
15        app:layout_constraintStart_toStartOf="parent"
16        app:layout_constraintTop_toTopOf="parent">
17        <ImageView
18            android:id="@+id/imageView"
19            android:layout_width="wrap_content"
20            android:layout_height="wrap_content"
21            app:srcCompat="@drawable/img2" />
22        <TextView
23            android:id="@+id/textView"
24            android:layout_width="wrap_content"
25            android:layout_height="wrap_content"
26            android:text=" 快乐大本营"
27            android:textColor="@android:color/black"
28            android:textSize="30sp" />
29        </FrameLayout>
30 </androidx.constraintlayout.widget.ConstraintLayout>
```

在上述代码中，灰色底纹的代码为系统自动生成，其他为拖进设计界面中组件后由系统生成的。程序运行结果如图 2.5 所示，布局文件中后添加的文本标签组件 TextView 遮挡了之前的图像组件 ImageView。

图 2.5　帧布局示例

4. 表格布局

表格布局（TableLayout）是将页面划分成行列构成的单元格。在 XML 布局文件中，由根元素 TableLayout 来标识表格布局。

表格的列数由 android:shrinkColumns 定义。例如，android:shrinkColumns = "1, 2, 3"，即表格为 3 列，其列编号为第 1、2、3 列。

表格的行数由<TableRow> </TableRow>定义。组件放置到哪一列由 android:layout_column 指定列编号。

【例 2-4】表格布局应用示例。设计一个 3 行 4 列的表格布局，组件安排如图 2.6 所示。

创建名称为 ex2_4 的新项目，包名为 com.example.ex2_4。生成项目框架后，将准备好的图像文件 img1.png～img5.png 复制到 res\drawable-hdpi 目录下。

在图 2.6 的界面布局中有未显示图片的空白区域，这时，可以使用空白组件 Space，这样显示出来的就是空白的单元格的形式。

组件在表格布局的结构如图 2.7 所示。

图 2.6　3 行 4 列的表格布局

图 2.7　表格布局的结构

界面布局文件 activity_main.xml 代码如下：

```
1   <?xml version="1.0" encoding="utf-8"?>
2   <androidx.constraintlayout.widget.ConstraintLayout
3     xmlns:android="http://schemas.android.com/apk/res/android"
4     xmlns:app="http://schemas.android.com/apk/res-auto"
5     xmlns:tools="http://schemas.android.com/tools"
6     android:layout_width="match_parent"
7     android:layout_height="match_parent"
8     tools:context=".MainActivity">

9   <TableLayout
10    android:layout_width="match_parent"
11    android:layout_height="match_parent"
12    android:shrinkColumns="0,1,2,3"          ◄── 设置表格为 4 列
13    tools:layout_editor_absoluteX="130dp"
14    tools:layout_editor_absoluteY="117dp">
15    <TableRow>    <!-- 第 1 行 -->
16      <ImageView  android:id="@+id/mImageView1"   ◄── 第 1 列
17        android:layout_width="wrap_content "
18        android:layout_height="wrap_content"
19        android:src="@drawable/img1" />                        ◄── 第 1 行
20      <ImageView  android:id="@+id/mImageView2"   ◄── 第 2 列
21        android:layout_width="wrap_content "
22        android:layout_height="wrap_content"
23        android:src="@drawable/img2" />
24    </TableRow>
```

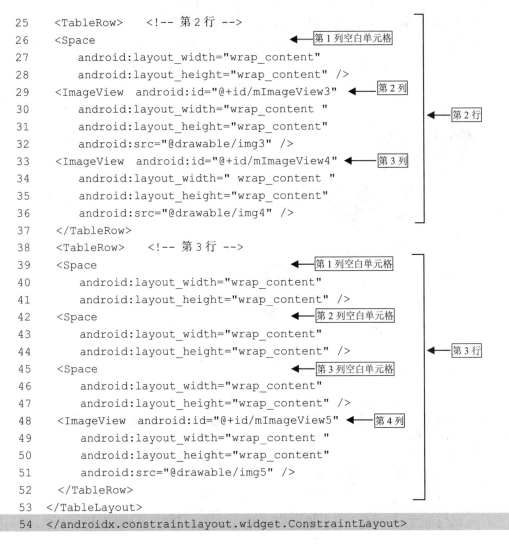

在上述代码中，灰色底纹的代码为系统自动生成，其他为拖进设计界面中组件后由系统生成的。

5．网格布局

网格布局（GridLayout）是把设置区域划分为若干行和若干列的网格，与 TableLayout 类似，在网格的内部放置各种需要的组件。

应用网格布局的属性可以设置组件在网格中的大小和摆放方式，网格中的一个组件可以占据多行或多列。

网格布局的主要属性如下：

- alignmentMode：设置布局管理器的对齐方式。
- columnCount：设置网格列数量。
- rowCount：设置网格行数量。
- layout_columnSpan：设置组件占据列数。

- layout_rowSpan：设置组件占据行数。

【例 2-5】 应用网格布局设计一个计算器界面。

计算器的设计界面如图 2.8 所示。在界面设计区域中，设置一个 6 行 4 列的网格布局，第 1 行为显示数据的文本标签；第 2 行为清除数据的按钮；第 3~6 行均分为 4 列，共安排 16 个按钮，分别代表数字 0、1、2、…、9 及加、减、乘、除、等号等。

图 2.8 应用网格布局设计计算器的界面

修改布局文件 activity_main.xml 如下：

```
1   <?xml version="1.0" encoding="utf-8"?>
2   <androidx.constraintlayout.widget.ConstraintLayout
3       xmlns:android="http://schemas.android.com/apk/res/android"
4       xmlns:app="http://schemas.android.com/apk/res-auto"
5       xmlns:tools="http://schemas.android.com/tools"
6       android:layout_width="match_parent"
7       android:layout_height="match_parent"
8       tools:context=".MainActivity">
9       <android.gridlayout.widget.GridLayout
10          android:id="@+id/gridLayout"
11          android:layout_width="0dp"
12          android:layout_height="0dp"
13          app:rowCount="6"                        设置网格为6行4列
14          app:columnCount="4"
15          app:layout_constraintBottom_toBottomOf="parent"
16          app:layout_constraintEnd_toEndOf="parent"
17          app:layout_constraintStart_toStartOf="parent"
18          app:layout_constraintTop_toTopOf="parent"
19          >
20          <!--文本标签-->
21          <TextView
22              android:layout_width="match_parent"
```

```
23      android:layout_height="wrap_content"
24      app:layout_columnSpan="4"      ◄── 该组件占据 4 列的位置
25      android:layout_marginLeft="4px"
26      android:gravity="left"
27      android:text="0"
28      android:textSize="50sp"
29    />
30    <Button
31      android:layout_width="match_parent"
32      android:layout_height="wrap_content"
33      app:layout_columnSpan="4"      ◄── 该组件占据 4 列的位置
34      android:text="清除"
35      android:textSize="26sp" />
36    <Button  android:text="1"  android:textSize="26sp" />
37    <Button  android:text="2"  android:textSize="26sp" />
38    <Button  android:text="3"  android:textSize="26sp" />
39    <Button  android:text="+"  android:textSize="26sp" />
40    <Button  android:text="4"  android:textSize="26sp" />
41    <Button  android:text="5"  android:textSize="26sp" />
42    <Button  android:text="6"  android:textSize="26sp" />
43    <Button  android:text="-"  android:textSize="26sp" />
44    <Button  android:text="7"  android:textSize="26sp" />
45    <Button  android:text="8"  android:textSize="26sp" />
46    <Button  android:text="9"  android:textSize="26sp" />
47    <Button  android:text="*"  android:textSize="26sp" />
48    <Button  android:text="."  android:textSize="26sp" />
49    <Button  android:text="0"  android:textSize="26sp" />
50    <Button  android:text="="  android:textSize="26sp" />
51    <Button  android:text="/"  android:textSize="26sp" />
52    </android.gridlayout.widget.GridLayout>
53  </androidx.constraintlayout.widget.ConstraintLayout>
```

在上述代码中，灰色底纹的代码为系统自动生成。

2.3 文本标签和按钮

2.3.1 文本标签

文本标签（TextView）用于显示文本内容，是最常用的组件之一。该组件常用的方法如表 2-2 所示，常用的 XML 文件元素属性如表 2-3 所示。

第2章 Android用户界面设计

表 2-2 文本标签（TextView）的常用方法

方　　法	功　　能
getText();	获取文本标签的文本内容
setText(CharSequence text);	设置文本标签的文本内容
setTextSize(float);	设置文本标签的文本大小
setTextColor(int color);	设置文本标签的文本颜色

表 2-3 文本标签（TextView）常用的 XML 文件元素

元 素 属 性	说　　明
android:id	文本标签标识
android:layout_width	文本标签 TextView 的宽度，通常取值 fill_parent（文本宽度）或以 pt 为单位的固定值
android:layout_height	文本标签 TextView 的高度，通常取值 wrap_content（文本高度）或以 px 为单位的固定值
android:text	文本标签 TextView 的文本内容
android:textSize	文本标签 TextView 的文本大小

视频讲解

【例 2-6】 设计一个文本标签组件程序。

创建名称为 ex2_6 的新项目，包名为 com.example.ex2_6。打开系统自动生成的项目框架，需要设计的文件如下：
- 设计资源文件 strings.xml；
- 设计布局文件 activity_main.xml；
- 设计控制文件 MainActivity.java。

（1）设计资源文件 strings.xml。

打开 res\values 下的 strings.xml，添加属性为"hello"的元素项的文本内容。

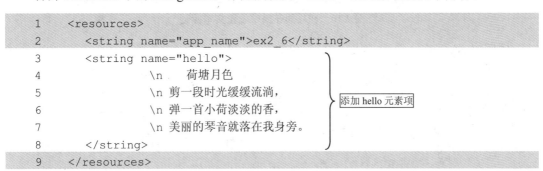

上述代码中，灰色底纹的代码为系统自动生成。

设计资源文件 strings.xml 的过程如图 2.9 所示。

（2）设计界面布局文件 activity_main.xml。

在界面布局文件 activity_main.xml 中加入文本标签组件 TextView，设置文本标签组件的 text 属性值为资源文件 strings.xml 中的 hello 项"@string/hello"，如图 2.10 所示。

图 2.9 添加属性为 "hello" 的元素项的文本内容

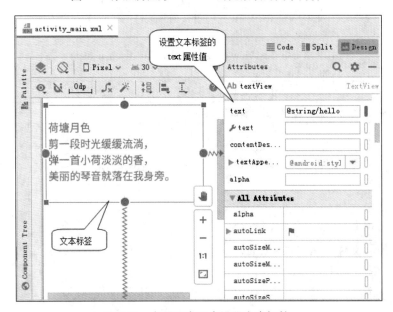

图 2.10 在界面布局中设置文本标签

完整的 activity_main.xml 代码如下：

```
1   <?xml version="1.0" encoding="utf-8"?>
2   <androidx.constraintlayout.widget.ConstraintLayout
3   xmlns:android="http://schemas.android.com/apk/res/android"
4       xmlns:app="http://schemas.android.com/apk/res-auto"
5       xmlns:tools="http://schemas.android.com/tools"
6       android:layout_width="match_parent"
7       android:layout_height="match_parent"
8       tools:context=".MainActivity">

9       <TextView
10          android:id="@+id/textView"
11          android:layout_width="357dp"
12          android:layout_height="213dp"
13          android:layout_marginStart="8dp"
14          android:layout_marginLeft="8dp"
```

```
15          android:layout_marginTop="32dp"
16          android:text="@string/hello"
17          android:textSize="24sp"/>

18   </androidx.constraintlayout.widget.ConstraintLayout>
```

在上述代码中，灰色底纹的代码为系统自动生成，<TextView>节点的代码为设置文本标签属性后由系统生成。

（3）设计控制文件 MainActivity.java。

在控制文件 MainActivity.java 源文件中添加文本标签组件，并将布局文件中所定义的文本标签元素属性值赋值给文本标签，与布局文件中文本标签建立关联。源程序如下：

```
1    package com.example.test2_1;
2    import androidx.appcompat.app.AppCompatActivity;
3    import android.os.Bundle;
4    import android.widget.TextView;           //引用文本标签组件

5    public class MainActivity extends Activity
6    {
7        private TextView  txt;                ◀── 声明文本标签对象
8        public void onCreate(Bundle savedInstanceState)
9        {
10           super.onCreate(savedInstanceState);
11           setContentView(R.layout.activity_main);
12           txt = (TextView)findViewById(R.id.textView1);   ◀── 与布局文件文本标签建立关联
13       }
14   }
```

在上述代码中，灰色底纹的代码为系统自动生成，其他代码为手动输入。保存项目，配置应用程序的运行参数，程序运行结果如图 2.11 所示。

图 2.11　文本标签

视频讲解

2.3.2 按钮及按钮处理事件

按钮（Button）用于处理人机交互事件，在一般应用程序中常常会用到。由于按钮（Button）是文本标签（TextView）的子类，其继承关系如图 2.12 所示。按钮（Button）继承了文本标签（TextView）的所有方法和属性。

```
java.lang.Object
   └─ android.view.View
        └─ android.widget.TextView
             └─ android.widget.Button
```

图 2.12 按钮（**Button**）与文本标签（**TextView**）的继承关系

按钮（Button）在程序设计中常用的方式是实现 OnClickListener 监听接口，当单击按钮时，通过 OnClickListener 监听接口触发 onClick()事件，实现用户需要的功能。OnClickListener 接口有一个 onClick()方法，在按钮（Button）实现 OnClickListener 接口时，一定要重写这个方法。

实现 OnClickListener 接口的类如下：

```
class mClick implements View.OnClickListener{
    public void onClick(View v)
    {
        //按钮处理事件代码
    }
}
```

按钮（Button）调用 OnClickListener 接口对象的方法如下：

```
按钮对象.setOnClickListener(new mClick);
```

【例 2-7】 编写程序，在点击按钮命令后，实现页面标题及文本组件的文字内容发生变化，如图 2.13 所示。

（点击按钮前）

（点击按钮后）

图 2.13 点击按钮后，文本组件的文字内容发生变化

创建名称为 ex2_7 的新项目，包名为 com.example.ex2_7。

(1) 设计布局文件 activity_main.xml。

在界面布局中添加一个文本标签,其 id 设置为 textView1,再添加一个按钮,其 id 设置为 button1,建立组件约束如图 2.14 所示。

完整的 activity_main.xml 文件代码如下:

图 2.14 建立文本标签和按钮组件的约束

```
1   <?xml version="1.0" encoding="utf-8"?>
2   <androidx.constraintlayout.widget.ConstraintLayout
3       xmlns:android="http://schemas.android.com/apk/res/android"
4       xmlns:app="http://schemas.android.com/apk/res-auto"
5       xmlns:tools="http://schemas.android.com/tools"
6       android:layout_width="match_parent"
7       android:layout_height="match_parent"
8       tools:context=".MainActivity">
9       <TextView
10          android:id="@+id/textView1"
11          android:layout_width="wrap_content"
12          android:layout_height="wrap_content"
13          android:layout_marginStart="60dp"
14          android:layout_marginLeft="60dp"
15          android:layout_marginTop="16dp"
16          android:text="Hello World!"
17          app:layout_constraintStart_toStartOf="parent"
18          app:layout_constraintTop_toTopOf="parent" />
19      <Button
20          android:id="@+id/button1"
21          android:layout_width="wrap_content"
22          android:layout_height="wrap_content"
23          android:text="点击我!"
24          android:layout_marginTop="20dp"
25          app:layout_constraintEnd_toEndOf="@+id/textView1"
26          app:layout_constraintHorizontal_bias="0.47"
27          app:layout_constraintStart_toStartOf="@+id/textView1"
28          app:layout_constraintTop_toBottomOf="@+id/textView1" />
29  </androidx.constraintlayout.widget.ConstraintLayout>
```

在上述代码中,灰色底纹的代码为系统自动生成,其他代码为设置组件约束及属性值后生成。

(2) 设计资源文件 strings.xml。

```
1   <resources>
2       <string name="app_name">ex2_2</string>
3       <string name="newStr">改变了文本标签的内容</string>
4   </resources>
```

(3) 设计控制文件 MainActivity.java。

在控制文件 MainActivity.java 中，设计一个实现按钮监听接口的内部类 mClick，当点击按钮时，触发 onClick()事件。

```java
1   package com.example.ex2_7;
2   import androidx.appcompat.app.AppCompatActivity;
3   import android.os.Bundle;
4   import android.view.View;
5   import android.view.View.OnClickListener;
6   import android.widget.TextView;
7   import android.widget.Button;
8   public class MainActivity extends AppCompatActivity
9   {
10      private TextView txt;              ← 声明与界面布局文件相同组件
11      private Button btn;
12      public void onCreate(Bundle savedInstanceState)
13      {
14          super.onCreate(savedInstanceState);
15          setContentView(R.layout.activity_main);
16          txt = (TextView) findViewById(R.id.textView1);   ← 与布局文件相关组件建立关联
17          btn = (Button)findViewById(R.id.button1);
18          btn.setOnClickListener(new mClick());            ← 注册监听接口
19      }
20      class mClick implements View.OnClickListener         ← 定义实现监听接口的内部类
21      {
22          public void onClick(View v)
23          {
24              MainActivity.this.setTitle("改变标题");
25              txt.setText(R.string.newStr);
26          }
27      }
28  }
```

在上述代码中，灰色底纹的代码为系统自动生成，其他代码为手动输入。

视频讲解

2.4 文本编辑框

```
android.view.View
  └─ android.widget.TextView
       └─ android.widget.EditText
```

图 2.15 文本编辑框（EditText）的继承关系

文本编辑框（EditText）用于接收用户输入的文本信息内容。文本编辑框（EditText）继承于文本标签（TextView），其继承关系如图 2.15 所示。

文本编辑框（EditText）主要继承文本标签（TextView）的方法，其常用的方法如表 2-4 所示。

表 2-4　文本编辑框（EditText）的常用方法

方　　法	功　　能
EditText(Context context)	构造方法，创建文本编辑框对象
getText()	获取文本编辑框的文本内容
setText(CharSequence text)	设置文本编辑框的文本内容

【例 2-8】 设计一个密码验证程序，输入的密码正确，显示"欢迎进入快乐大本营！"，否则显示"非法用户，请立刻离开！"，如图 2.16 所示。

图 2.16　密码验证

创建名称为 ex2_8 的新项目，包名为 com.example.ex2_8。
（1）设计布局文件 activity_main.xml。

在界面布局中，设置一个编辑框组件 EditText（选择组件面板中的 Plain Text 选项），用于输入密码；再设置一个按钮组件 Button，判断密码是否正确；设置两个文本标签组件 TextView，其中一个显示提示信息"请输入密码："，另一个用于显示密码正确与否。

各组件的属性值如表 2-5 所示。

表 2-5　各组件的属性值

组件 id	属　　性	属　性　值
TextView1	text	请输入密码：
EditText	text	（空）
Button	text	确定
TextView2	text	（空）

建立各组件约束如图 2.17 所示。

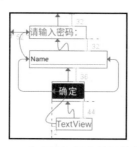

图 2.17　密码验证程序的组件约束

界面布局文件 activity_main.xml 程序代码如下：

```
1   <?xml version="1.0" encoding="utf-8"?>
2   <androidx.constraintlayout.widget.ConstraintLayout
3       xmlns:android="http://schemas.android.com/apk/res/android"
4       xmlns:app="http://schemas.android.com/apk/res-auto"
5       xmlns:tools="http://schemas.android.com/tools"
6       android:layout_width="match_parent"
7       android:layout_height="match_parent"
8       tools:context=".MainActivity">
9       <!--建立一个 TextView -->
10      <TextView
11          android:id="@+id/textView1"
12          android:layout_width="fill_parent"
13          android:layout_height="41px"
14          android:layout_x="33px"
15          android:layout_y="106px"
16          android:text="请输入密码:"
17          android:textSize="24sp"
18          />
19      <!--建立一个 EditText -->
20      <EditText
21          android:id="@+id/editText"
22          android:layout_width="180px"
23          android:layout_height="wrap_content"
24          android:layout_x="29px"
25          android:layout_y="33px"
26          android:inputType="text"
27          android:textSize="24sp" />
28      <!--建立一个 Button -->
29      <Button
30          android:id="@+id/button"
31          android:layout_width="100px"
32          android:layout_height="wrap_content"
33          android:text="确定"
34          android:textSize="24sp"
35          />
36      <!--建立一个 TextView -->
37      <TextView
38          android:id="@+id/textView2"
39          android:layout_width="180px"
40          android:layout_height="41px"
41          android:layout_x="33px"
42          android:layout_y="106px"
43          android:textSize="24sp"
44          />
45  </androidx.constraintlayout.widget.ConstraintLayout>
```

（2）设计控制文件 MainActivity.java。

在控制文件 MainActivity.java 中，主要是设计按钮的监听事件，当单击按钮后，从文本编辑框中获取输入的文本内容，与密码"abc123"进行比较。

```java
1   package com.example.ex2_8;
2   import androidx.appcompat.app.AppCompatActivity;
3   import android.os.Bundle;
4   import android.view.View;
5   import android.view.View.OnClickListener;
6   import android.widget.EditText;
7   import android.widget.TextView;
8   import android.widget.Button;

9   public class MainActivity extends AppCompatActivity
10  {
11      private EditText edit;              ← 声明与界面布局程
12      private TextView txt1,txt2;            序中相同的组件
13      private Button  btn;
14      @Override
15      public void onCreate(Bundle savedInstanceState)
16      {
17          super.onCreate(savedInstanceState);
18          setContentView(R.layout.activity_main);
19          txt1 = (TextView)findViewById(R.id.textView1);  ← 与界面布局程
20          txt2 = (TextView)findViewById(R.id.textView2);     序中的组件建
21          edit = (EditText)findViewById(R.id.editText);      立关联
22          btn = (Button)findViewById(R.id.button);
23          btn.setOnClickListener(new mClick());  ← 设置监听对象
24      }

25      class mClick implements OnClickListener  ← 定义实现监听接口的内部类
26      {
27          public void onClick(View v)
28          {
29              String passwd;
30              passwd=edit.getText().toString();  ← 获取文本编辑框中的文本内容
31              if(passwd.equals("abc123"))  ← 用equals()方法比较两个字符串是否相等
32                  txt2.setText("欢迎进入快乐大本营!");
33              else
34                  txt2.setText("非法用户,请立刻离开!");
35          }
36      }
37  }
```

2.5 进度条和选项按钮

2.5.1 进度条

视频讲解

进度条 ProgressBar、SeekBar 和 RatingBar 都能以图示方式直观地显示某个过程的进度。

1. ProgressBar 组件

进度条 ProgressBar 的常用属性和方法如表 2-6 所示。

表 2-6 进度条 ProgressBar 的常用属性和方法

属 性	方 法	功 能
android:max	setMax(int max)	设置进度条的变化范围为 0～max
android:progress	setProgress(int progress)	设置进度条的当前值（初始值）
android:progress	incrementProgressBy(int diff)	进度条的变化步长值

2. SeekBar 组件

组件 SeekBar 是 ProgressBar 的子类，继承了 ProgressBar 的全部属性和方法。通过使用 SeekBar 组件，用户可方便地拖曳进度条的滑块，从而改变过程的参数值。

在设计 SeekBar 组件程序时要实现 OnSeekBarChangeListener 监听接口，并重写该接口的 3 个监听方法。

```
public void onProgressChanged(SeekBar seekBar, int progress, boolean fromUser)
{    } // 进度条的值改变时发生

public void onStartTrackingTouch(SeekBar seekBar)
{    } // 进度条开始拖曳时发生

public void onStopTrackingTouch(SeekBar seekBar)
{    } // 进度条停止拖曳时发生
```

3. RatingBar 组件

组件 RatingBar 也是 ProgressBar 的子类，继承了 ProgressBar 的全部属性和方法。该组件通过五角星图案表示进度，主要用于等级评分。

RatingBar 组件的主要属性如下。

- android:numStars：显示多少个星星，必须为整数。
- android:rating：默认评分值，必须为浮点数。
- android:stepSize：评分每次增加的值，必须为浮点数。

【例 2-9】 进度条应用示例。

（1）设计界面。

在界面设计中，分别设置 ProgressBar 组件、SeekBar 组件和 RatingBar 组件，并设置两个按钮，用于控制进度条的进度变化，如图 2.18 所示。

组件的属性设置如表 2-7 所示。

表 2-7　各组件的属性设置

组件 id	属　性	属　性　值	组件 id	属　性	属　性　值
ProgressBar	max	100	Button1	text	增加
SeekBar	max	100	Button2	text	减少
RatingBar	—	—			

在界面布局文件 activity_main.xml 中，建立各组件的约束如图 2.19 所示。

图 2.18　进度条进度控制

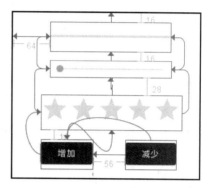

图 2.19　进度条应用的组件约束

（2）设计控制程序 MainActivity.java。

```
1   package com.example.ex2_9;
2   import androidx.appcompat.app.AppCompatActivity;
3   import android.os.Bundle;
4   import android.view.View;
5   import android.widget.Button;
6   import android.widget.ProgressBar;
7   import android.widget.RatingBar;
8   import android.widget.SeekBar;

9   public class MainActivity extends AppCompatActivity
10  {
11      ProgressBar pb;
12      SeekBar sb;
13      RatingBar rb;
14      Button btn1,btn2;
15      @Override
16      public void onCreate(Bundle savedInstanceState)
17      {
18          super.onCreate(savedInstanceState);
```

```
19      setContentView(R.layout.activity_main);
20      pb = (ProgressBar)findViewById(R.id.progressBar);
21      sb = (SeekBar)findViewById(R.id.seekBar);
22      rb = (RatingBar)findViewById(R.id.ratingBar);
23      btn1=(Button)findViewById(R.id.button1);
24      btn2=(Button)findViewById(R.id.button2);
25      btn1.setOnClickListener(new click1());
26      btn2.setOnClickListener(new click2());
27      sb.setOnSeekBarChangeListener(new mSeekBar());
28    }

29    class click1 implements View.OnClickListener
30    {
31      public void onClick(View v)
32      { sb.incrementProgressBy(10); }
33    }

34    class click2 implements View.OnClickListener
35    {
36      public void onClick(View v)
37      { sb.incrementProgressBy(-10); }
38    }

39    class mSeekBar implements SeekBar.OnSeekBarChangeListener
40    {
41      @Override
42      public void onProgressChanged(
43          SeekBar seekBar, int progress, boolean fromUser)
44      {
45        pb.setProgress(sb.getProgress());
46        rb.setRating((sb.getProgress())/20);
47      }
48      @Override
49      public void onStartTrackingTouch(SeekBar seekBar) { }
50      @Override
51      public void onStopTrackingTouch(SeekBar seekBar) { }
52    }

53 }
```

(第19~24行：与用户界面程序中的组件建立关联)
(第32行：进度增加)
(第37行：进度减少)
(第39~52行：监听SeekBar拖曳)

2.5.2 选项按钮

1. 复选按钮

复选按钮（CheckBox）用于多项选择的情形，用户可以一次性选择多个选项。复选按钮（CheckBox）是按钮（Button）的子类，其属性与方法继承按钮（Button）。复选按钮

（CheckBox）常用方法如表 2-8 所示。

表 2-8　复选按钮（CheckBox）的常用方法

方　　法	功　　能
isChecked()	判断选项是否被选中
getText()	获取复选按钮的文本内容

【例 2-10】　复选按钮应用示例。

在界面设计中，安排 3 个复选按钮和 1 个普通按钮，选择选项后点击按钮，在文本标签中显示所选中的选项文本内容，如图 2.20 所示。

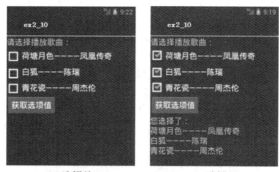

(a) 选择前　　　　　　(b) 选择后

图 2.20　复选按钮的多项选择

程序设计步骤：

（1）在布局文件中，首先设置一个垂直线性布局组件 LinearLayout（vertical），然后再设置 3 个复选按钮 CheckBox。

（2）在 strings.xml 文件中设置要使用的字符串。

（3）在 MainActivity.java 文件中，声明与用户界面程序中相同的组件并建立它们之间的关联。调用 CheckBox 的 isChecked()方法判断该选项是否被选中，如果被选中，则调用 getText()方法获取选项的文本内容。

程序代码设计：

（1）设计布局文件 activity_main.xml。

```
1   <?xml version="1.0" encoding="utf-8"?>
2   <androidx.constraintlayout.widget.ConstraintLayout
3       xmlns:android= "http://schemas.android.com/apk/res/android"
4       xmlns:app="http://schemas.android.com/apk/res-auto"
5       xmlns:tools="http://schemas.android.com/tools"
6       android:layout_width="match_parent"
7       android:layout_height="match_parent"
8       tools:context=".MainActivity">
9   <LinearLayout
10      android:id="@+id/linearLayout"
```

```xml
11      android:layout_width="0dp"
12      android:layout_height="0dp"
13      android:orientation="vertical"
14      app:layout_constraintBottom_toBottomOf="parent"
15      app:layout_constraintEnd_toEndOf="parent"
16      app:layout_constraintStart_toStartOf="parent"
17      app:layout_constraintTop_toTopOf="parent">

18      <TextView
19        android:layout_width="fill_parent"
20        android:layout_height="wrap_content"
21        android:text="@string/hello"
22        android:textSize="20sp"/>
23      <CheckBox
24        android:id="@+id/check1"
25        android:layout_width="fill_parent"
26        android:layout_height="wrap_content"
27        android:textSize="20sp"
28        android:text="@string/one" />
29      <CheckBox
30        android:id="@+id/check2"
31        android:layout_width="fill_parent"
32        android:layout_height="wrap_content"
33        android:textSize="20sp"
34        android:text="@string/two" />
35      <CheckBox
36        android:id="@+id/check3"
37        android:layout_width="fill_parent"
38        android:layout_height="wrap_content"
39        android:textSize="20sp"
40        android:text="@string/three" />
41      <Button
42        android:id="@+id/button"
43        android:layout_width="wrap_content"
44        android:layout_height="wrap_content"
45        android:textSize="20sp"
46        android:text="@string/btn" />
47      <TextView
48        android:id="@+id/textView2"
49        android:layout_width="fill_parent"
50        android:layout_height="wrap_content"
51        android:text=""
52        android:textSize="20sp"/>
53      </LinearLayout>

54  </androidx.constraintlayout.widget.ConstraintLayout>
```

(2) 在 strings.xml 文件中设置要使用的字符串。

```
1   <resources>
2       <string name="app_name">ex2_10</string>
3       <string name="hello">请选择播放歌曲：</string>
4       <string name="one">荷塘月色————凤凰传奇</string>
5       <string name="two">白狐————陈瑞</string>
6       <string name="three">青花瓷————周杰伦</string>
7       <string name="btn">获取选项值</string>
8   </resources>
```

(3) 设计控制程序 MainActivity.java。

在控制程序 MainActivity.java 中，建立程序中组件与用户界面程序组件的关联，并编写按钮的事件处理代码。

```
1   package com.example.ex2_10;

2   import androidx.appcompat.app.AppCompatActivity;
3   import android.os.Bundle;

4   import android.view.View;
5   import android.view.View.OnClickListener;
6   import android.widget.Button;
7   import android.widget.CheckBox;
8   import android.widget.TextView;

9   public class MainActivity extends AppCompatActivity
10  {
11      CheckBox ch1,ch2,ch3;
12      Button okBtn;
13      TextView txt;

14      @Override
15      public void onCreate(Bundle savedInstanceState)
16      {
17          super.onCreate(savedInstanceState);
18          setContentView(R.layout.activity_main);

19          ch1=(CheckBox)findViewById(R.id.check1);
20          ch2=(CheckBox)findViewById(R.id.check2);
21          ch3=(CheckBox)findViewById(R.id.check3);
22          okBtn=(Button)findViewById(R.id.button);
23          txt=(TextView)findViewById(R.id.textView2);
24          okBtn.setOnClickListener(new click());
25      }

26      class click implements View.OnClickListener
```

与用户界面程序中的组件建立关联

```
27      {
28           public void onClick(View v)
29           {
30               String str="";
31               if(ch1.isChecked()) str=str+"\n"+ch1.getText();
32               if(ch2.isChecked()) str=str+"\n"+ch2.getText();
33               if(ch3.isChecked()) str=str+"\n"+ch3.getText();
34               txt.setText("您选择了: "+str);
35           }
36      }
37 }
```

2. 单选组件与单选按钮

单选组件（RadioGroup）用于多项选择中只允许任选其中一项的情形。单选组件（RadioGroup）由一组单选按钮（RadioButton）组成。单选按钮（RadioButton）是按钮（Button）的子类。单选按钮（RadioButton）的常用方法如表 2-9 所示。

表 2-9 单选按钮（RadioButton）的常用方法

方　法	功　能
isChecked();	判断选项是否被选中
getText();	获取单选按钮的文本内容

【例 2-11】 单选按钮应用示例。

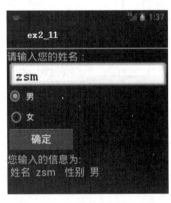

图 2.21 单选按钮应用示例

在界面设计中，设置两个单选按钮、一个文本编辑框和一个普通按钮，选择选项后点击按钮，在文本标签中显示文本编辑框及所选中的选项文本内容，如图 2.21 所示。

程序设计步骤：

（1）在界面布局文件中，首先设置一个垂直线性布局组件 LinearLayout（vertical），然后设置两个单选按钮 RadioButton，并将单选按钮的 text 属性值分别设置为"男"和"女"。

（2）在 Activity 中，声明与用户界面程序中相同的组件并建立它们之间的关联。

（3）调用 RadioButton 的 isChecked()方法判断该选项是否被选中，如果被选中，则调用 getText()方法获取选项的文本内容。

程序代码设计：

（1）设计布局文件 activity_main.xml。

```
1   <?xml version="1.0" encoding="utf-8"?>
2   <androidx.constraintlayout.widget.ConstraintLayout
3       xmlns:android="http://schemas.android.com/apk/res/android"
```

```
4      xmlns:app="http://schemas.android.com/apk/res-auto"
5      xmlns:tools="http://schemas.android.com/tools"
6      android:layout_width="match_parent"
7      android:layout_height="match_parent"
8      tools:context=".MainActivity">

9      <LinearLayout
10        android:id="@+id/linearLayout"
11        android:layout_width="0dp"
12        android:layout_height="0dp"
13        android:orientation="vertical"
14        app:layout_constraintBottom_toBottomOf="parent"
15        app:layout_constraintEnd_toEndOf="parent"
16        app:layout_constraintStart_toStartOf="parent"
17        app:layout_constraintTop_toTopOf="parent">
18        <TextView
19         android:layout_width="wrap_content"
20         android:layout_height="wrap_content"
21         android:text="请输入您的姓名："
22         android:textSize="20sp" />
23        <EditText
24         android:layout_width="match_parent"
25         android:layout_height="wrap_content"
26         android:id="@+id/editText" />
27        <RadioButton
28         android:layout_width="wrap_content"
29         android:layout_height="wrap_content"
30         android:text="男"
31         android:id="@+id/radioButton1"
32         android:textSize="20sp" />
33        <RadioButton
34         android:layout_width="wrap_content"
35         android:layout_height="wrap_content"
36         android:text="女"
37         android:id="@+id/radioButton2"
38         android:textSize="20sp" />
39        <Button
40         android:layout_width="wrap_content"
41         android:layout_height="wrap_content"
42         android:text="确定"
43         android:id="@+id/button"
44         android:textSize="20sp" />
45        <TextView
46         android:layout_width="fill_parent"
47         android:layout_height="wrap_content"
48         android:id="@+id/textView2"
49         android:textSize="20sp" />
```

```
50        </LinearLayout>
```

```
51    </androidx.constraintlayout.widget.ConstraintLayout>
```

（2）设计控制程序 MainActivity.java。

在控制程序 MainActivity.java 中，建立程序中组件与用户界面程序组件的关联，并编写按钮的事件处理代码。

```
1   package com.example.ex2_11;
2   import androidx.appcompat.app.AppCompatActivity;
3   import android.os.Bundle;
4   import android.view.View;
5   import android.widget.Button;
6   import android.widget.EditText;
7   import android.widget.TextView;
8   import android.widget.RadioButton;

9   public class MainActivity extends AppCompatActivity
10  {
11      RadioButton r1,r2;
12      Button okBtn;
13      TextView txt;
14      EditText edit;
15      @Override
16      public void onCreate(Bundle savedInstanceState)
17      {
18          super.onCreate(savedInstanceState);
19          setContentView(R.layout.activity_main);
20          edit = (EditText)findViewById(R.id.editText);
21          edit = (EditText)findViewById(R.id.editText)
22          txt =(TextView)findViewById(R.id.textView2);         ┐
23          okBtn = (Button)findViewById(R.id.button);            │ 与用户界面程
24          r1 = (RadioButton)findViewById(R.id.radioButton1);    │ 序中的组件建
25          r2 = (RadioButton)findViewById(R.id.radioButton2);   ┘ 立关联
26          okBtn.setOnClickListener(new mClick());
27      }
28      class mClick implements OnClickListener
29      {
30          public void onClick(View v)
31          {
32              String str = "", name = "";
33              name = edit.getText();
34              if (r1.isChecked())      ← 第1个单选按钮被选中
35                  str = r1.getText();
36              if (r2.isChecked())      ← 第2个单选按钮被选中
37                  str = r2.getText();
```

```
38              txt.setText("您输入的信息为: \n 姓名 " + name + "\t 性别 " + str);
39          }
40      }
41  }
```

2.6 图像显示类

视频讲解

图像显示类(ImageView)用于显示图片或图标等图像资源,并提供图像缩放及着色(渲染)等图像处理功能。

ImageView 类的常用属性和对应方法如表 2-10 所示。

表 2-10 ImageView 类的常用属性和对应方法

属　　性	对 应 方 法	说　　明
android:maxHeight	setMaxHeight(int)	为显示图像提供最大高度的可选参数
android:maxWidth	setMaxWidth(int)	为显示图像提供最大宽度的可选参数
android:scaleType	setScaleType(ImageView.ScaleType)	控制图像适合 ImageView 大小的显示方式(见表 2-13)
android:src	setImageResource(int)	获取图像文件的路径

ImageView 类的 scaleType 属性值如表 2-11 所示。

表 2-11 ImageView 类的 scaleType 属性值

属性值常量	值	说　　明
matrix	0	用矩阵来绘图
fitXY	1	拉伸图像(不按宽高比例)以填充 View 的宽高
fitStart	2	按比例拉伸图像,拉伸后图像的高度为 View 的高度,且显示在 View 的左边
fitCenter	3	按比例拉伸图像,拉伸后图像的高度为 View 的高度,且显示在 View 的中间
fitEnd	4	按比例拉伸图像,拉伸后图像的高度为 View 的高度,且显示在 View 的右边
center	5	按原图大小显示图像,当图像宽高大于 View 的宽高时,截取图像中间部分显示
centerCrop	6	按比例放大原图直至等于某边 View 的宽高显示
centerInside	7	当原图宽高等于 View 的宽高时,按原图大小居中显示;否则,将原图缩放至 View 的宽高居中显示

【例 2-12】 图像显示应用实例。

程序设计步骤:

(1)将事先准备好的多张图片序列 img1.jpg～img6.jpg 复制到 res\drawable 目录下。

(2)在界面布局文件中设置图像显示组件 ImageView。

(3)在 Activity 中获得相关组件实例。

（4）通过触发按钮事件调用 OnClickListener 接口的 onClick()方法显示图像。

程序代码设计：

（1）设计用户界面程序 activity_main.xml。

在界面设计中，安排两个按钮和一个图像显示组件 ImageView，点击按钮可以翻阅浏览图片。图像显示的布局设计如图 2.22 所示。

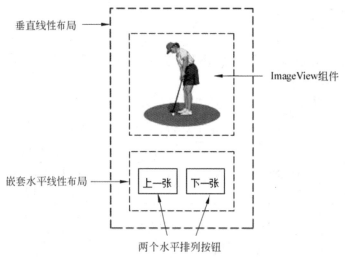

图 2.22　图像显示的布局设计

安排好组件后需要设置图像显示组件的数据源，其步骤如下：

① 选取 ImageView 组件，在属性窗口中选择 srcCompat 选项，点击右边的按钮设置显示图像的数据源，如图 2.23 所示。

图 2.23　设置显示图像的数据源

② 在弹出的 Pick a Resource 对话框中，选择 Drawable 选项，选取复制在 res\drawable

目录下的图像，如图 2.24 所示。

图 2.24　选取复制在 res\drawable 目录下的图像

用户界面程序 activity_main.xml 的代码如下：

```xml
1   <?xml version="1.0" encoding="utf-8"?>
2   <androidx.constraintlayout.widget.ConstraintLayout
3       xmlns:android="http://schemas.android.com/apk/res/android"
4       xmlns:app="http://schemas.android.com/apk/res-auto"
5       xmlns:tools="http://schemas.android.com/tools"
6       android:layout_width="match_parent"
7       android:layout_height="match_parent"
8       tools:context=".MainActivity">
9       <LinearLayout
10          android:id="@+id/linearLayout"
11          android:layout_width="0dp"
12          android:layout_height="0dp"
13          android:orientation="vertical"
14          app:layout_constraintBottom_toBottomOf="parent"
15          app:layout_constraintEnd_toEndOf="parent"
16          app:layout_constraintStart_toStartOf="parent"
17          app:layout_constraintTop_toTopOf="parent">
18          <ImageView
19              android:id="@+id/imageView"
20              android:layout_width="match_parent"
```

```
21          android:layout_height="350dp"
22          app:srcCompat="@drawable/img1" />
23      <LinearLayout
24          android:layout_width="fill_parent"
25          android:layout_height="wrap_content"
26          android:gravity="center_horizontal"
27          android:orientation="horizontal" >
28          <Button
29              android:id="@+id/btn_last"
30              android:layout_width="150dp"
31              android:layout_height="wrap_content"
32              android:text="上一张" />
33          <Button
34              android:id="@+id/btn_next"
35              android:layout_width="150dp"
36              android:layout_height="wrap_content"
37              android:text="下一张"  />
38      </LinearLayout>
39  </LinearLayout>
40 </androidx.constraintlayout.widget.ConstraintLayout>
```

（2）设计控制程序 MainActivity.java。

在控制程序 MainActivity.java 中，建立程序中组件与用户界面程序组件的关联，并编写按钮的事件处理代码。

```
1   package com.example.ex2_12;
2   import androidx.appcompat.app.AppCompatActivity;
3   import android.os.Bundle;
4   import android.view.View;
5   import android.widget.Button;
6   import android.widget. EditText;
7   import android.widget. TextView;
8   import android.widget.RadioButton;
9   public class MainActivity extends AppCompatActivity
10  {
11      ImageView img;
12      Button btn_last, btn_next;
13      //存放图像id的int类型数组
14      private int[] imgs = {
15          R.drawable.img1,
16          R.drawable.img2,
17          R.drawable.img3,
18          R.drawable.img4,
19          R.drawable.img5,
20          R.drawable.img6 };
```

数组元素为资源目录中的图像序列 img1.jpg～img6.jpg

```
21      int index = 0;       //数组元素的位置
22    @Override
23    public void onCreate(Bundle savedInstanceState)
24    {
25      super.onCreate(savedInstanceState);
26      setContentView(R.layout.activity_main);
27      img = (ImageView)findViewById(R.id.img);              ─┐ 与用户界面程序中
28      btn_last = (Button)findViewById(R.id.btn_last);        │ 的组件进行关联
29      btn_next = (Button)findViewById(R.id.btn_next);       ─┘
30      btn_last.setOnClickListener(new mClick());            ─┐ 注册监听接口
31      btn_next.setOnClickListener(new mClick());            ─┘
32    }

33    class mClick implements OnClickListener    //定义一个类实现监听接口
34    {
35      public void onClick(View v)
36      {
37        if(v == btn_last)
38        {
39          if(index>0 && index<imgs.length)        ─┐
40          {                                        │
41            index--;                               │ "上一张"按钮事件
42            img.setImageResource(imgs[index]);     │
43          } else {index = imgs.length+1; }        ─┘
44        }
45        if(v == btn_next)
46        {
47          if(index>0&&index<imgs.length-1)         ─┐
48          {                                        │
49            index++;                               │ "下一张"按钮事件
50            img.setImageResource(imgs[index]);     │
51          }else {index = imgs.length-1;   }       ─┘
52        }
53      }
54    }
55  }
```

程序运行结果如图 2.25 所示。

图 2.25　图像显示示例

2.7 消息提示类

在 Android 系统中，可以用消息提示类（Toast）来显示帮助或提示消息。提示消息以浮于应用程序之上的形式显示在屏幕上，因为它并不获得焦点，所以不会影响用户的其他操作。使用消息提示组件 Toast 的目的是为了尽可能不中断用户操作，并使用户看到提供的信息内容。Toast 类的常用方法及说明如表 2-12 所示。

表 2-12　Toast 类的常用方法及说明

方　　法	说　　明
Toast(Context context)	Toast 的构造方法，构造一个空的 Toast 对象
makeText(Context context, CharSequence text, int duration)	以特定时长显示文本内容，参数 text 为显示的文本，参数 duration 为显示的时间，较长时间取值 LENGTH_LONG、较短时间取值 LENGTH_SHORT
setDuration(int duration)	设置存续时间
setView(View view)	设置要显示的视图
setGravity(int gravity, int xOffset, int yOffset)	设置提示信息在屏幕上的显示位置
setText(int resId)	更新 makeText()方法所设置的文本内容
show( )	显示提示信息
LENGTH_LONG	提示信息显示较长时间的常量
LENGTH_SHORT	提示信息显示较短时间的常量

【例 2-13】 消息提示 Toast 分别按默认方式、自定义方式显示的示例。

将事先准备好的图标文件 icon.jpg 复制到资源目录 res\ drawable 目录下，以备提示消息的图标之用。

（1）设计布局文件 activity_main.xml。

在界面设计中，设置一个文本标签和两个按钮，分别对应消息提示 Toast 的两种显示方式，程序代码如下。

```xml
1  <?xml version="1.0" encoding="utf-8"?>
2  <androidx.constraintlayout.widget.ConstraintLayout
3      xmlns:android="http://schemas.android.com/apk/res/android"
4      xmlns:app="http://schemas.android.com/apk/res-auto"
5      xmlns:tools="http://schemas.android.com/tools"
6      android:layout_width="match_parent"
7      android:layout_height="match_parent"
8      tools:context=".MainActivity">

9      <LinearLayout
10         android:id="@+id/linearLayout"
11         android:layout_width="0dp"
12         android:layout_height="0dp"
13         android:orientation="vertical"
14         app:layout_constraintBottom_toBottomOf="parent"
15         app:layout_constraintEnd_toEndOf="parent"
16         app:layout_constraintStart_toStartOf="parent"
17         app:layout_constraintTop_toTopOf="parent">

18         <TextView
19             android:layout_width="match_parent"
20             android:layout_height="wrap_content"
21             android:gravity="center"    ← 居中显示文本
22             android:text="消息提示 Toast"
23             android:textSize="24sp" />
24         <Button
25             android:id="@+id/btn1"
26             android:layout_height="wrap_content"
27             android:layout_width="match_parent"
28             android:text="默认方式"
29             android:textSize="20sp" />
30         <Button
31             android:id="@+id/btn2"
32             android:layout_height="wrap_content"
33             android:layout_width="match_parent"
34             android:text="自定义方式"
35             android:textSize="20sp" />
36     </LinearLayout>

37 </androidx.constraintlayout.widget.ConstraintLayout>
```

（2）设计事件处理文件 MainActivity.java。

```
1   package com.example.ex2_13;
2   import androidx.appcompat.app.AppCompatActivity;
3   import android.os.Bundle;
4   import android.view.View;
5   import android.widget.Button;
6   import android.widget. EditText;
7   import android.widget. TextView;
8   import android.widget.RadioButton;
9   public class MainActivity extends AppCompatActivity
10  {
11      ListView list;
12      Button btn1,btn2;

13      @Override
14      public void onCreate(Bundle savedInstanceState)
15      {
16          super.onCreate(savedInstanceState);
17          setContentView(R.layout.activity_main);
18          btn1=(Button)findViewById(R.id.btn1);
19          btn2=(Button)findViewById(R.id.btn2);
20          btn1.setOnClickListener(new mClick());        ← 为按钮注册事件监听器
21          btn2.setOnClickListener(new mClick());
22      }

23      class mClick implements OnClickListener
24      {
25          Toast toast;
26          @Override
27          public void onClick(View v)
28          {
29              if(v==btn1)        ← 居中显示文本
30              {
31                  Toast.makeText(getApplicationContext(),   ← 设置提示消息内容,可用
32                          "默认Toast方式",                      MainActivity.this 替换
33                  Toast.LENGTH_SHORT).show();                getApplicationContext()
34              }
35              else if(v==btn2)
36              {
37                  toast = Toast.makeText(getApplicationContext(),
38                          "自定义Toast的位置",
39                          Toast.LENGTH_SHORT);
40                  toast.setGravity(Gravity.CENTER, 0, 0);   ← 自定义显示位置
41                  toast.show();
42              }
43          }
44      }
45  }
```

程序运行结果如图 2.26 所示。

图 2.26 消息提示示例

2.8 列表组件类

列表组件类（ListView）是 Android 程序开发中经常用到的组件，该组件必须与适配器配合使用，由适配器提供显示样式和显示数据。

ListView 类的常用方法及说明如表 2-13 所示。

表 2-13 ListView 类的常用方法及说明

方 法	说 明
ListView(Context context)	构造方法
addHeaderView(View v)	设置列表项目的头部
addFooterView(View v)	设置列表项目的底部
setAdapter(ListAdapter adapter)	为列表设置显示数组选项的适配器
setOnItemClickListener(AdapterView.OnItemClickListener listener)	注册单击选项时执行的方法，该方法继承于父类 android.widget.AdapterView

【例 2-14】 列表组件应用示例。

在界面设计中，设置一个文本标签和一个列表组件 ListView，程序代码如下。

（1）设计布局文件 activity_main.xml。

```
1    <?xml version="1.0" encoding="utf-8"?>
2    <androidx.constraintlayout.widget.ConstraintLayout
3        xmlns:android="http://schemas.android.com/apk/res/android"
4        xmlns:app="http://schemas.android.com/apk/res-auto"
5        xmlns:tools="http://schemas.android.com/tools"
```

```
6        android:layout_width="match_parent"
7        android:layout_height="match_parent"
8        tools:context=".MainActivity">

9    <LinearLayout
10       android:id="@+id/linearLayout"
11       android:layout_width="0dp"
12       android:layout_height="0dp"
13       android:orientation="vertical"
14       app:layout_constraintBottom_toBottomOf="parent"
15       app:layout_constraintEnd_toEndOf="parent"
16       app:layout_constraintStart_toStartOf="parent"
17       app:layout_constraintTop_toTopOf="parent">

18       <TextView
19           android: id="@+id/textView"
20           android:layout_width=" match_parent"
21           android:layout_height=" match_parent"
22           android:text="凤凰传奇"
23           android:textSize="24sp" />
24       <ListView
25           android:id="@+id/listView"
26           android:layout_height="match_parent"
27           android:layout_width=" match_parent" />
28   </LinearLayout>

29 </androidx.constraintlayout.widget.ConstraintLayout>
```

（2）设计事件处理文件 MainActivity.java。

```
1  package com.example.ex2_14;
2  import androidx.appcompat.app.AppCompatActivity;
3  import android.os.Bundle;
4  import android.view.View;
5  import android.widget.AdapterView;
6  import android.widget.ArrayAdapter;
7  import android.widget.ListView;
8  import android.widget.TextView;
9  import android.widget.Toast;

10 public class MainActivity extends AppCompatActivity
11 {
12     ListView list;
13     String[] data ={           //定义数组
14         "（1）荷塘月色",
15         "（2）最炫民族风",
16         "（3）天蓝蓝",
```

```
17          "(4)最美天下",
18          "(5)自由飞翔",
19      };

20      @Override
21      public void onCreate(Bundle savedInstanceState)
22      {
23          super.onCreate(savedInstanceState);
24          setContentView(R.layout.activity_main);
25          list= (ListView)findViewById(R.id.listView);    ← 与界面的列表组件建立关联
26          //为 ListView 设置数组适配器 ArrayAdapter
27          list.setAdapter(new ArrayAdapter<String>(
28              this,
29              android.R.layout.simple_expandable_list_item_1,    ← 为列表设置数组
30              data ));                                              适配器和监听器
31          //为 ListView 设置列表选项监听器
32          list.setOnItemClickListener(new mItemClick());
33      }


34      //定义列表选项监听器的事件
35      class mItemClick implements AdapterView.OnItemClickListener
36      {
37          @Override
38          public void onItemClick(
39                  AdapterView<?> arg0,
40                  View arg1,
41                  int arg2,
42                  long arg3)
43          {
44              Toast.makeText(
45                  MainActivity.this,
46                  "您选择的项目是: " + ((TextView)view).getText(),    ← 提示信息
47                  Toast.LENGTH_SHORT
48              ).show();
49          }
50      }

51      }
```

上述代码的相关语句说明:

(1) android.R.layout.simple_list_item_1 是 Android 系统内置的 ListView 布局方式。

- android.R.layout.simple_list_item_1:设置一行 text。
- android.R.layout.simple_list_item_2:设置一行 title,一行 text。
- android.R.layout.simple_list_item_single_choice:设置单选按钮。
- android.R.layout.simple_list_item_multiple_choice:设置多选按钮。

（2）OnItemClickListener 是一个接口，用于监听列表组件选项的触发事件。

（3）Toast.makeText().show() 显示提示消息。

程序运行结果如图 2.27 所示。

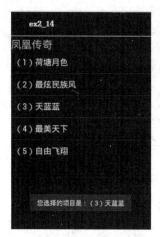

图 2.27　列表组件示例

习题 2

1. 编写程序，实现单击按钮将文本编辑框中的文字内容显示到文本标签，如图 2.28 所示。

图 2.28　文本编辑框中的文字内容显示到文本标签

2. 设计一个加法计算器，如图 2.29 所示，在前两个文本编辑框中输入整数，当单击按钮"＝"时，在第 3 个文本编辑框中显示这两个数之和。

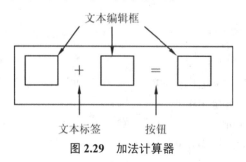

图 2.29　加法计算器

3. 设计如图 2.30 所示的用户界面。

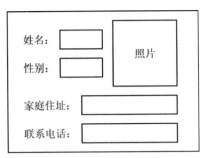

图 2.30　设计用户界面

4. 完成例 2-5 计算器的 MainActivity.java 控制程序代码设计。

5. 编制一个"我的故乡"图册，下方配有说明文字，单击"上一张""下一张"按钮，图片切换时说明文字也随之切换，且"上一张""下一张"按钮的下方显示前一张及后一张的图片缩略图，如图 2.31 所示。

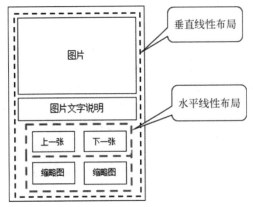

图 2.31　浏览图片及缩略图

第3章 多个用户界面的程序设计

3.1 页面切换与传递参数值

3.1.1 绑定机制组件

Intent 是 Android 系统的一种运行时的绑定机制,在应用程序运行时连接两个不同组件。在 Android 的应用程序中,不管是页面切换、传递数据,还是调用外部程序,都可能要用到 Intent。Intent 负责对应用中某次操作的动作、动作涉及的数据、附加数据进行描述。Android 则根据 Intent 的描述,负责找到对应的组件,将 Intent 传递给调用的组件,并完成组件的调用。因此,可以将 Intent 理解为不同组件之间通信的"媒介",其专门提供组件互相调用的相关信息。

Intent 的属性有动作(Action)、数据(Data)、分类(Category)、类型(Type)、组件(Component)以及扩展(Extra),其中最常用的是 Action 属性。

例如:

Intent.ACTION_MAIN	表示标识 Activity 为一个程序的开始。
Intent.ACTION_GET_CONTENT	表示允许用户选择图片或录音等特殊种类的数据。
Intent.ACTION_SMS_SEND	表示发送邮件的动作。
Intent.ACTION_SMS_RECEIVED	表示接收邮件的动作。
Intent.ACTION_ANSWER	表示处理呼入的电话。
Intent.ACTION_CALL_BUTTON	表示按"拨号"键。
Intent.ACTION_CALL	表示呼叫指定的电话号码。

3.1.2 Activity 页面切换

Activity 跳转与传递参数值主要通过 Intent 类协助实现。在一个 Activity 页面中启动另一个 Activity 页面的运行,这是最简单的 Activity 页面切换方式。

页面切换的步骤如下:

(1) 首先创建一个 Intent 对象,其构造方法如下。

```
Intent intent = new Intent(当前Activity.this, 另一Activity.class);
```
（2）调用 Activity 的 startActivity(intent)方法，切换到另一个 Activity 页面。

【例 3-1】 从一个 Activity 页面启动另一个 Activity 页面示例。

创建名称为 ex3_1 的新项目，包名为 com.example.ex3_1。在本项目中要建立两个页面文件及两个控制文件，第 1 个页面的界面布局文件为 activity_main.xml、控制文件为 MainActivity.java，第 2 个页面的界面布局文件为 activity_second.xml、控制文件为 secondActivity.java。此外，还要修改配置文件 AndroidManifest.xml。

（1）设计第 1 个页面。

① 进入系统自动生成的应用程序框架，在页面的界面布局中，设置一个文本标签组件和一个按钮组件，建立组件约束如图 3.1 所示。

图 3.1 第 1 个页面的界面布局及组件约束

② 修改第 1 个页面的控制文件 MainActivity.java，源代码如下：

```
1   package com.example.ex3_1;
2   import androidx.appcompat.app.AppCompatActivity;
3   import android.os.Bundle;
4   import android.widget.Button;
5   import android.content.Intent;
6   import android.view.View;

7   public class MainActivity extends AppCompatActivity
8   {
9       private Button okBtn;
10      Intent intent;
11  @Override
12  public void onCreate(Bundle savedInstanceState)
13  {
14      super.onCreate(savedInstanceState);
15      setContentView(R.layout.activity_main);
16      btn = (Button)findViewById(R.id.button);
17      btn.setOnClickListener(new mClick());;
18  }
19  class btnclock implements OnClickListener    ← 定义实现监听接口的类
20  {
21      public void onClick(View v)
22      {
23          intent = new Intent(MainActivity.this, secondActivity.class);
```

```
24            startActivity(intent);    //启动新的Activity页面
25        }
26    }
27 }
```

(2)设计第2个页面。

① 右击项目管理器中的 app/java/com.example.ex3_1 选项，在弹出的快捷菜单中选择 New（新建）→Activity→Empty Activity 选项，如图 3.2 所示。

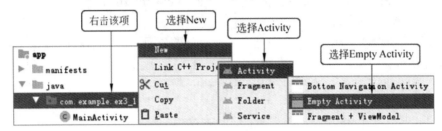

图 3.2　新建第 2 个页面

在弹出的对话框中，输入第 2 个页面的 Activity 名称 SecondActivity，其页面布局 Layout 名称为 activity_second，如图 3.3 所示。

图 3.3　配置第 2 个页面

这时，系统自动生成第 2 个页面的界面布局文件 activity_second.xml 和控制文件 secondActivity.java。

② 在第 2 个页面的界面中，设置一个文本标签，将其 text 属性值设置为"这是第 2 个

页面"。自动生成的第 2 个页面的控制程序 secondActivity.java 代码不需要任何修改,其代码如下。

```
1  ackage com.example.ex3_1;
2  import androidx.appcompat.app.AppCompatActivity;
3  import android.os.Bundle;
4  ublic class SecondActivity extends AppCompatActivity {
5      @Override
6      protected void onCreate(Bundle savedInstanceState) {
7          super.onCreate(savedInstanceState);
8          setContentView(R.layout.activity_second);
9      }
10 }
```

(3) 配置文件 AndroidManifest.xml。

打开项目中的 AndroidManifest.xml 配置文件,可以看到增加了一行注册第 2 个 Activity 页面的代码,其程序代码如下。

```
1  <?xml version="1.0" encoding="utf-8"?>
2  <manifest xmlns:android="http://schemas.android.com/apk/res/android"
3      package="com.example.ex3_1">
4      <application
5          android:allowBackup="true"
6          android:icon="@mipmap/ic_launcher"
7          android:label="@string/app_name"
8          android:roundIcon="@mipmap/ic_launcher_round"
9          android:supportsRtl="true"
10         android:theme="@style/Theme.Ex3_1">
11         <activity android:name=".SecondActivity"></activity>   ← 新增的第 2 个页面的注册语句
12         <activity android:name=".MainActivity">
13             <intent-filter>
14                 <action android:name="android.intent.action.MAIN" />
15                 <category android:name="android.intent.category.LAUNCHER" />
16             </intent-filter>
17         </activity>
18     </application>
19 </manifest>
```

程序运行结果如图 3.4 所示。

图 3.4　从一个页面切换到另一页面

3.1.3 在 Activity 页面之间传递数据

1. Bundle 类

Bundle 类是用于将字符串与某组件对象建立映射关系的组件。Bundle 组件与 Intent 配合使用，可以在不同的 Activity 之间传递数据。Bundle 类的常用方法如下。
- putString(String key, String value)：把字符串用"键-值"对形式存放到 Bundle 对象中。
- remove(String key)：移除指定 key 的值。
- getString(String key)：获取指定 key 的字符。

2. Intent 操作 Bundle 组件的方法

Intent 操作绑定组件 Bundle 的方法如下。
- getExtras()：获取 Intent 组件中绑定的 Bundle 对象。
- putExtras()：把 Bundle 对象绑定到 Intent 组件上。

3. 应用 Intent 在不同的 Activity 之间传递数据

下面说明应用 Intent 与 Bundle 配合从一个 Activity 页面传递数据到另一 Activity 页面的方法。

（1）在页面 Activity A 端。

① 创建 Intent 对象和 Bundle 对象：

```
Intent intent = new Intent();
Bundle bundle = new Bundle();
```

② 为 Intent 指定切换页面，用 Bundle 存放"键-值"对数据：

```
intent.setClass(MainActivity.this, SecondActivity.class);
bundle.putString("text", txt.getText().toString());
```

③ 将 Bundle 对象传递给 Intent：

```
intent.putExtras(bundle);
```

（2）在另一页面 Activity B 端。

① 从 Intent 中获取 Bundle 对象：

```
bunde = this.getIntent().getExtras();
```

② 从 Bundle 对象中按"键-值"对的键名获取对应数据值：

```
String str = bunde.getString("text");
```

在不同的 Activity 页面之间传递数据的过程如图 3.5 所示。

第3章 多个用户界面的程序设计

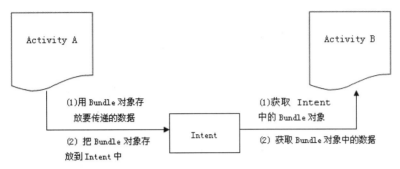

图 3.5 应用 Intent 在 Activity 页面之间传递数据

【例 3-2】 从第 1 个 Activity 页面传递数据到第 2 个 Activity 页面示例。

（1）设计第 1 个页面的界面布局 activity_main.xml。

在第 1 个页面的界面布局中，设置一个文本标签和一个按钮，建立组件约束如图 3.6 所示。

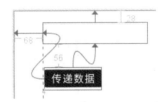

图 3.6 建立第 1 个页面的界面布局和组件约束

设置编辑框 EditText 的 background 属性值：

android:background = "@android:drawable/editbox_background"。

（2）设计第 1 个页面的控制程序 MainActivity.java。

```
1   package com.example.ex3_2;
2   import androidx.appcompat.app.AppCompatActivity;
3   import android.content.Intent;
4   import android.os.Bundle;
5   import android.view.View;
6   import android.widget.Button;
7   import android.widget.EditText;

8   public class MainActivity extends AppCompatActivity
9   {
10      EditText txt1;
11      Button   btn1;
12      @Override
13      protected void onCreate(Bundle savedInstanceState)
14      {
15          super.onCreate(savedInstanceState);
16          setContentView(R.layout.activity_main);
17          txt1 = (EditText)findViewById(R.id.editText);
```

```
18        btn1 = (Button)findViewById(R.id.button1);
19        btn1.setOnClickListener(new mClick());
20    }

21    class mClick implements View.OnClickListener{
22     @Override
23     public void onClick(View v) {
24      Intent intent = new Intent();                               设置 Intent 对
25      intent.setClass(MainActivity.this, SecondActivity.class);   象切换的页面
26      Bundle bundle = new Bundle();
27      bundle.putString("text", txt1.getText().toString());        Bundle 对象
28      intent.putExtras(bundle);                                   绑定数据
29      startActivity(intent);      启动另一个 Activity 页面
30     }
31    }
32 }
```

（3）设计第 2 个页面的界面布局文件 activity_second.xml。

在项目中生成第 2 个页面，在页面中设置一个文本标签和一个按钮，其界面布局如图 3.7 所示。

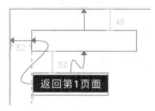

图 3.7　建立第 2 个页面的界面布局和组件约束

（4）设计第 2 个页面的控制文件 secondActivity.java，其代码如下。

```
1  package com.example.ex3_2;
2  import androidx.appcompat.app.AppCompatActivity;
3  import android.content.Intent;
4  import android.os.Bundle;
5  import android.view.View;
6  import android.widget.Button;
7  import android.widget.EditText;

8  public class MainActivity extends AppCompatActivity
9  {
10    TextView txt2;
11    Button  btn2;
12    Intent intent2;
13    Bundle bundle2;
14    @Override
15    protected void onCreate(Bundle savedInstanceState)
16    {
```

```
17        super.onCreate(savedInstanceState);
18        setContentView(R.layout.activity_main);
19        txt2 = (TextView)findViewById(R.id.textView);
20        btn2 = (Button)findViewById(R.id.button2);
21        intent2 = this.getIntent();
22        bundle2 = intent2.getExtras();
23        String str = bundle2.getString("text");   ← 获取键名为 text 的值
24        txt2.setText(str);
25        btn2.setOnClickListener(new mClick());
26    }

27    //定义返回到前一页面的监听接口事件
28    class btnclock2 implements OnClickListener
29    {
30      public void onClick(View v)
31      {
32        Intent intent3 = new Intent();
33        intent3.setClass(SecondActivity.this, MainActivity.class);
34        startActivity(intent3);   ← 返回前一页面
35      }
36    }
37 }
```

运行程序，在第 1 个页面的编辑框中输入数据，点击按钮后跳转到第 2 个页面，数据也随之传递到第 2 个页面，如图 3.8 所示。

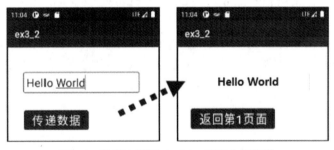

图 3.8　数据在不同 Activity 页面之间传递

3.2 菜单设计

一个菜单（Menu）由多个菜单选项组成，选择一个菜单项就可以引发一个动作事件。

在 Android 系统中，菜单可以分为 3 类：选项菜单（Option Menu）、上下文菜单（Context Menu）和子菜单（Sub Menu）。下面主要介绍选项菜单和上下文菜单的设计方法，由于子菜单的设计方法基本与选项菜单相同，这里就不赘述了。

3.2.1 选项菜单

选项菜单需要通过按下设备的 Menu 键来显示。当按下设备上的 Menu 键后，在屏幕底部会弹出一个菜单，这个菜单称为选项菜单（Option Menu）。

1. Activity 中创建菜单的方法

设计选项菜单需要用到 Activity 中的 onCreateOptionMenu(Menu menu)方法，用于建立菜单并且在菜单中添加菜单项；还需要用到 Activity 中的 onOptionsItemSelected(MenuItem item)方法，用于响应菜单事件。Activity 实现选项菜单的方法如表 3-1 所示。

表 3-1 Activity 实现选项菜单的方法

方 法	说 明
onCreateOptionMenu(Menu menu)	用于初始化菜单，menu 为 Menu 对象实例
onPrepareOptionsMenu(Menu menu)	改变菜单状态，在菜单显示前调用
onOptionsMenuClosed(Menu menu)	菜单被关闭时调用
onOptionsItemSelected(MenuItem item)	菜单项被点击时调用，即菜单项的监听方法

2. 菜单 Menu

设计选项菜单需要用到 Menu、MenuItem 接口。一个 Menu 对象代表一个菜单，在 Menu 对象中可以添加菜单项 MenuItem 对象，也可以添加子菜单 Sub Menu。

菜单 Menu 使用 add(int groupId, int itemId, int order, CharSequence title) 方法添加一个菜单项，add()方法中的 4 个参数如下。

（1）groupId（组别）：如果不分组就写 Menu.NONE。

（2）itemId（id）：很重要，Android 根据这个 id 来确定不同的菜单。

（3）order（顺序）：哪个菜单项在前面由这个参数的大小决定。

（4）title（标题）：菜单项的显示文本。

3. 创建选项菜单的步骤

创建选项菜单的步骤如下：

（1）重写 Activity 的 onCreateOptionMenu(Menu menu)方法，当菜单第一次被打开时调用。

（2）调用 Menu 的 add( )方法添加菜单项（MenuItem）。

（3）重写 Activity 的 onOptionsItemSelected(MenuItem item)方法，当菜单项（MenuItem）被选择时来响应事件。

【例 3-3】 选项菜单应用示例。

设计一个选项菜单应用的示例程序，其运行结果如图 3.9 所示。

（1）设计界面布局文件 activity_main.xml。

在界面布局文件中设置一个文本标签，用于显示选择的菜单项。

图 3.9 菜单示例

（2）设计事件处理的控制程序 MainActivity.java。

控制程序 MainActivity.java 的源代码如下：

```
1    package com.example.ex3_3;
2    import androidx.appcompat.app.AppCompatActivity;
3    import android.os.Bundle;

4    import android.view.Menu;
5    import android.view.MenuItem;
6    import android.widget.TextView;

7    public class MainActivity extends AppCompatActivity
8    {
9        TextView txt;
10   @Override
11   public void onCreate(Bundle savedInstanceState)
12   {
13       super.onCreate(savedInstanceState);
14       setContentView(R.layout.activity_main);
15       txt = (TextView)findViewById(R.id.TextView1);
16   }

17   public boolean onCreateOptionsMenu(Menu menu)
18   {
19       // 调用父类方法来加入系统菜单
20       super.onCreateOptionsMenu(menu);
21       // 添加菜单项
22       menu.add(
23               1,          //组号
24               1,          //唯一的 id
25               1,          //序号
26               "菜单项1"); //菜单项标题
27       menu.add( 1, 2, 2,  "菜单项2");
28       menu.add( 1, 3, 3,  "菜单项3");
29       menu.add( 1, 4, 4,  "菜单项4");
```

添加菜单项的 4 个参数

```
30          return true;
31      }

32      public boolean onOptionsItemSelected(MenuItem item)
33      {
34        String title = "选择了" + item.getTitle().toString();
35        switch (item.getItemId())
36        { //响应每个菜单项(通过菜单项的id)
37          case 1:
38              txt.setText(title);   ← 文本标签显示菜单项的标题
39              break;
40          case 2:
41              txt.setText(title);   ← 文本标签显示菜单项的标题
42              break;
43          case 3:
44              txt.setText(title);   ← 文本标签显示菜单项的标题
45              break;
46          case 4:
47              txt.setText(title);   ← 文本标签显示菜单项的标题
48              break;
49           default:
50              //没有处理的事件交给父类来处理
51              return super.onOptionsItemSelected(item);
52        }
53        return true;
54      }
55  }
```

3.2.2 上下文菜单

Android 系统中的上下文菜单类似于计算机上的右键菜单。在为一个视图注册了上下文菜单之后，长按（两秒左右）这个视图对象就会弹出一个浮动菜单，即上下文菜单。任何视图都可以注册上下文菜单，最常见的是用于列表视图 ListView 的 item。

创建一个上下文菜单的步骤如下：

（1）重写 Activity 的 onCreateContenxtMenu()方法，调用 Menu 的 add 方法添加菜单项（MenuItem）。

（2）重写 Activity 的 onContextItemSelected()方法,响应上下文菜单菜单项的单击事件。

（3）调用 Activity 的 registerForContextMenu()方法，为视图注册上下文菜单。

【例 3-4】 上下文菜单应用示例。

设计一个上下文菜单应用的示例程序，其运行结果如图 3.10 所示。

（1）设计界面布局文件 activity_main.xml。

在界面布局文件中设置 3 个文本标签，用于显示选择项，建立其约束如图 3.11 所示。

第3章 多个用户界面的程序设计

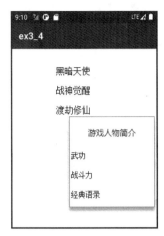

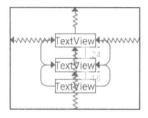

图 3.10 上下文菜单应用示例　　　　图 3.11 建立 3 个文本标签的约束

（2）设计事件处理的控制程序 MainActivity.java。

控制程序 MainActivity.java 的源代码如下：

```
1   package com.example.ex3_4;
2   import androidx.appcompat.app.AppCompatActivity;
3   import android.os.Bundle;
4   import android.view.Menu;
5   import android.view.MenuItem;
6   import android.view.ContextMenu;
7   import android.widget.TextView;
8   import android.view.View;
9   public class MainActivity extends AppCompatActivity
10  {
11      TextView txt1, txt2, txt3;
12      private static final int item1 = Menu.FIRST;
13      private static final int item2 = Menu.FIRST+1;
14      private static final int item3 = Menu.FIRST+2;
15      String str[] = {"黑暗天使","战神觉醒","渡劫修仙"};
16      @Override
17      public void onCreate(Bundle savedInstanceState)
18      {
19          super.onCreate(savedInstanceState);
20          setContentView(R.layout.activity_main);
21          txt1 = (TextView)findViewById(R.id.TextView1);
22          txt2 = (TextView)findViewById(R.id.textView2);
23          txt3 = (TextView)findViewById(R.id.textView3);
24          txt1.setText(str[0].toString());
25          txt2.setText(str[1].toString());
26          txt3.setText(str[2].toString());
27          registerForContextMenu(txt1);
28          registerForContextMenu(txt2);
29          registerForContextMenu(txt3);
```

← 初始化组件

75

```
30      }
31      //上下文菜单，本例会通过长按条目激活上下文菜单
32      @Override
33      public void onCreateContextMenu(ContextMenu menu, View view,
34      ContextMenuInfo menuInfo)
35      {
36          menu.setHeaderTitle("游戏人物简介");
37          //添加菜单项
38          menu.add(0, item1, 0, "武功");
39          menu.add(0, item2, 0, "战斗力");
40          menu.add(0, item3, 0, "经典语录");
41      }
42      //菜单单击响应
43      @Override
44      public boolean onContextItemSelected(MenuItem item)
45      {
46          //获取当前被选择的菜单项的信息
47          switch(item.getItemId())
48          {
49           case item1:
50              //在这里添加处理代码
51              break;
52           case item2:
53              //在这里添加处理代码
54              break;
55           case item3:
56              //在这里添加处理代码
57              break;
58          }
59          return true;
60      }
61  }
```

（选项的具体功能没有实现）

3.3 对话框

对话框是一个有边框、有标题栏的独立存在的容器，在应用程序中经常使用对话框组件来进行人机交互。Android 系统提供了 4 种常用对话框。

- AlertDialog：消息对话框。
- ProgressDialog：进度条对话框。
- DatePickerDialog：日期选择对话框。
- TimePickerDialog：时间选择对话框。

下面逐一介绍这些对话框的使用方法。

3.3.1 消息对话框

AlertDialog 对话框是应用程序设计中最常用的对话框之一。AlertDialog 对话框的内容很丰富,使用 AlertDialog 类可以创建普通对话框、带列表的对话框以及带单选按钮和多选按钮的对话框。AlertDialog 类的常用方法如表 3-2 所示。

表 3-2 AlertDialog 类的常用方法

方法	说明
AlertDialog.Builder(Context)	对话框 Builder 对象的构造方法
create();	创建 AlertDialog 对象
setTitle();	设置对话框标题
setIcon();	设置对话框图标
setMessage();	设置对话框的提示信息
setItems();	设置对话框要显示的一个 list
setPositiveButton();	在对话框中添加 yes 按钮
setNegativeButton();	在对话框中添加 no 按钮
show();	显示对话框
dismiss();	关闭对话框

创建 AlertDialog 对象需要使用 AlertDialog 的内部类 Builder。设计 AlertDialog 对话框的步骤如下。

(1) 用 AlertDialog.Builder 类创建对话框 Builder 对象。

```
Builder dialog=new AlertDialog.Builder(Context);
```

(2) 设置对话框的标题、图标、提示信息内容、按钮等。

```
dialog.setTitle("普通对话框");
dialog.setIcon(R.drawable.icon1);
dialog.setMessage("一个简单的提示对话框");
dialog.setPositiveButton("确定", new okClick());
```

(3) 创建并显示 AlertDialog 对话框对象。

```
dialog.create();
dialog.show();
```

如果在对话框内部设置了按钮,还需要为其设置事件监听 OnClickListener。

【例 3-5】 消息对话框应用示例。

在本例中设计了两种形式的对话框程序,一种是发出提示信息的普通对话框,另一种是用户登录对话框。

在用户登录对话框中,设计了用户登录的布局文件 long.xml,供用户输入相关验证信息。

在创建的应用程序框架中,将事先准备的图像文件 icon1.jpg、icon2.jpg 复制到 res\drawable 目录下,用作对话框的图标。

程序的运行结果如图 3.12 所示。

(a) 普通对话框　　　　　　　　(b) 用户登录对话框

图 3.12　AlertDialog 对话框

(1) 设计界面布局文件 activity_main.xml。

在界面中设置两个按钮,建立约束如图 3.13 所示。

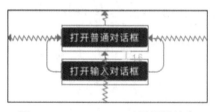

图 3.13　AlertDialog 对话框的约束

(2) 设计登录对话框的界面布局文件 login.xml。

新建一个 xml 文件,命名为 login.xml,其代码如下:

```
1    <?xml version="1.0" encoding="utf-8"?>
2    <LinearLayout xmlns:android="http://schemas.android.com/apk/res/android"
3        android:layout_width="match_parent"
4        android:layout_height="match_parent"
5        android:orientation="vertical" >
6        <TextView
7            android:id="@+id/user"
8            android:layout_width="wrap_content"
9            android:layout_height="wrap_content"
10           android:text="用户名"
11           android:textSize="18sp"/>
```

```
12    <EditText
13        android:id="@+id/userEdit"
14        android:layout_width="match_parent"
15        android:layout_height="wrap_content"
16        android:textSize="18sp"/>
17    <TextView
18        android:id="@+id/textView"
19        android:layout_width="match_parent"
20        android:layout_height="wrap_content"
21        android:text="密码"
22        android:textSize="18sp"/>
23    <EditText
24        android:id="@+id/paswdEdit"
25        android:layout_width="match_parent"
26        android:layout_height="wrap_content"
27        android:textSize="18sp"/>
28 </LinearLayout>
```

(3) 设计控制文件 MainActivity.java。

```
1  package com.example.ex3_5;
2  import androidx.appcompat.app.AppCompatActivity;
3  import android.os.Bundle;
4  import android.app.AlertDialog;
5  import android.content.DialogInterface;
6  import android.view.View;
7  import android.widget.Button;
8  import android. widget.EditText;
9  import android.widget.LinearLayout;
10 import android.widget.Toast;
11 public class MainActivity extends AppCompatActivity
12 {
13     Button btn1, btn2;
14     LinearLayout login;
15     @Override
16     public void onCreate(Bundle savedInstanceState)
17     {
18         super.onCreate(savedInstanceState);
19         setContentView(R.layout.activity_main);
20         btn1 = (Button)findViewById(R.id.button1);
21         btn2 = (Button)findViewById(R.id.button2);
22         btn1.setOnClickListener(new mClick());
23         btn2.setOnClickListener(new mClick());
24     }
25     class mClick implements OnClickListener
26     {
```

```
27        Builder dialog=new AlertDialog.Builder(MainActivity.this);
28        @Override
29         public void onClick(View v)
30        {
31          if(v == btn1)
32          {
33            //设置对话框的标题
34            dialog.setTitle("警告");
35            //设置对话框的图标
36            dialog.setIcon(R.drawable.icon1);
37            //设置对话框显示的内容
38            dialog.setMessage("本项操作可能导致信息泄露！");
39            //设置对话框的"确定"按钮
40            dialog.setPositiveButton("确定", new okClick());
41            //创建对象框
42            dialog.create();
43            //显示对象框
44            dialog.show();
45          }
46          else  if(v == btn2)
47          {
48            login = (LinearLayout)getLayoutInflater()
49                    .inflate(R.layout.login, null);
50            dialog.setTitle("用户登录").setMessage("请输入用户名和密码")
51                    .setView(login);
52            dialog.setPositiveButton("确定", new loginClick());
53            dialog.setNegativeButton("退出", new exitClick());
54            dialog.setIcon(R.drawable.icon2);
55            dialog.create();
56            dialog.show();
57          }
58        }
59     }
60     /*  普通对话框的"确定"按钮事件 */
61     class okClick implements DialogInterface.OnClickListener
62     {
63       @Override
64        public void onClick(DialogInterface dialog, int which)
65       {
66          dialog.cancel();    ← 关闭对话框
67       }
68     }
69     /*  输入对话框的"确定"按钮事件   */
70     class loginClick implements  DialogInterface.OnClickListener
71     {
72       EditText txt;
73        @Override
```

```
74     public void onClick(DialogInterface dialog, int which)
75     {
76       txt = (EditText)login.findViewById(R.id.paswdEdit);    ← 关联布局文件中的组件
77       //取出输入编辑框的值与密码"admin"比较
78       if((txt.getText().toString()).equals("admin"))
79         Toast.makeText(getApplicationContext(),
80             "登录成功", Toast.LENGTH_SHORT).show();            ← 密码为 admin 时,显示"登录成功"
81       else
82         Toast.makeText(getApplicationContext(),
83             "密码错误", Toast.LENGTH_SHORT).show();
84       dialog.dismiss();    ← 关闭对话框
85     }
86   }
87   /* 输入对话框的"退出"按钮事件  */
88   class exitClick implements DialogInterface.OnClickListener
89   {
90     @Override
91     public void onClick(DialogInterface dialog, int which)
92     {
93       MainActivity.this.finish();    ← 点击"退出"按钮,退出 MainActivity 程序
94     }
95   }
96 }
```

对于程序的第 48、49 行:

```
login = (LinearLayout)getLayoutInflater()
            .inflate(R.layout.login, null);
```

这里 inflate()是将组件从一个 XML 中定义的布局找出来。

在一个 Activity 中如果直接用 findViewById(),对应的是 setConentView()中的那个 layout 中的组件(程序第 19 行中的 R.layout.activity_main)。如果 Activity 中用到其他 layout 布局,比如对话框上的 layout,还要设置对话框上的 layout 中的组件(像图片 ImageView、文字 TextView)上的内容,这就必须用 inflate()先将对话框上的 layout 找出来,然后再用这个 layout 对象找到它上面的组件。

3.3.2 其他几种常用对话框

1. 进度条对话框

Android 系统有一个 ProgressDialog 类,它继承于 AlertDialog 类,综合了进度条与对话框的特点,使用起来非常简单。ProgressDialog 类的继承关系如图 3.14 所示。

```
java.lang.Object
  └ android.app.Dialog
      └ android.app.AlertDialog
          └ android.app.ProgressDialog
```

图 3.14 ProgressDialog 类继承于 AlertDialog 类

ProgressDialog 类的常用方法如表 3-3 所示。

表 3-3 ProgressDialog 类的常用方法

方　　法	说　　明
getMax()	获取对话框进度的最大值
getProgress()	获取对话框当前的进度值
onStart()	开始调用对话框
setMax(int max)	设置对话框进度的最大值
setMessage(CharSequence message)	设置对话框的文本内容
setProgress(int value)	设置对话框当前的进度
show(Context context, CharSequence title, CharSequence message)	设置对话框的显示内容和方式
ProgressDialog(Context context)	对话框的构造方法

2. 日期对话框和时间对话框

日期选择类 DatePickerDialog 和时间选择类 TimePickerDialog 都继承于 AlertDialog 类，一般用于日期和时间的设定，它们的常用方法如表 3-4 所示。

表 3-4 日期和时间选择对话框的常用方法

方　　法	说　　明
updateDate(int year, int monthOfYear, int dayOfMonth)	设置 DatePickerDialog 对象的当前日期
onDateChanged(DatePicker view, int year, int month, int day)	修改 DatePickerDialog 对象的日期
updateTime(int hourOfDay, int minutOfHour)	设置 TimePickerDialog 对象的时间
onTimeChanged(TimePicker view, int hourOfDay, int minute)	修改 TimePickerDialog 对象的时间

【例 3-6】 进度对话框、日期对话框和时间对话框示例。

（1）设计用户界面程序 activity_main.xml。

在界面设计中，设置 3 个按钮，分别用于打开进度对话框、日期对话框和时间对话框。

（2）设计控制程序 MainActivity.java。

```
1    package com.example.ex3_6;
2    import androidx.appcompat.app.AppCompatActivity;
3    import android.os.Bundle;

4    package com.example.ex3_6;
5    import android.app.Activity;
6    import android.app.DatePickerDialog;
7    import android.app.ProgressDialog;
8    import android.app.TimePickerDialog;
9    import android.app.DatePickerDialog.OnDateSetListener;
10   import android.app.TimePickerDialog.OnTimeSetListener;
11   import android.view.View;
12   import android.view.View.OnClickListener;
13   import android.widget.Button;
14   import android.widget.DatePicker;
15   import android.widget.TimePicker;
```

```java
16  public class MainActivity extends Activity
17  {
18      Button btn1,btn2,btn3;
19      @Override
20      public void onCreate(Bundle savedInstanceState)
21      {
22          super.onCreate(savedInstanceState);
23          setContentView(R.layout.activity_main);
24          btn1=(Button)findViewById(R.id.button1);
25          btn2=(Button)findViewById(R.id.button2);
26          btn3=(Button)findViewById(R.id.button3);
27          btn1.setOnClickListener(new mClick());
28          btn2.setOnClickListener(new mClick());
29          btn3.setOnClickListener(new mClick());
30      }
31      class mClick implements OnClickListener
32      {
33          int m_year = 2012;
34          int m_month = 1;
35          int m_day = 1;
36          int m_hour = 12,  m_minute = 1;
37          @Override
38          public void onClick(View v)
39          {
40              if(v == btn1)
41              {
42                  ProgressDialog  d=new ProgressDialog (MainActivity.this);
43                  d.setTitle("进度对话框");
44                  d.setIndeterminate(true);
45                  d.setMessage("程序正在 Loading...");
46                  d.setCancelable(true);
47                  d.setMax(10);
48                  d.show();
49              }
50              else  if(v == btn2)
51              {
52                  //设置日期监听器
53                  DatePickerDialog.OnDateSetListener dateListener =
54                          new DatePickerDialog.OnDateSetListener()
55                  {
56                      @Override
57                      public void onDateSet(DatePicker view, int year,
58                              int month, int dayOfMonth)
59                      {
60                          m_year = year;
61                          m_month = month;
```

```
62                    m_day = dayOfMonth;
63                }
64            };
65            //创建日期对话框对象
66            DatePickerDialog date = new DatePickerDialog(MainActivity.this,
67                    dateListener, m_year, m_month, m_day);
68            date.setTitle("日期对话框");
69            date.show();
70        }
71        else  if(v == btn3)
72        {    //设置时间监听器
73            TimePickerDialog.OnTimeSetListener timeListener =
74                    new TimePickerDialog.OnTimeSetListener()
75            {
76                @Override
77                public void onTimeSet(TimePicker view,
78                        int hourOfDay, int minute)
79                {
80                    m_hour = hourOfDay;
81                    m_minute = minute;
82                }
83            };
84            TimePickerDialog d = new TimePickerDialog(MainActivity.this,
85                    timeListener, m_hour, m_minute, true);
86            d.setTitle("时间对话框");
87            d.show();
88        }
89       }
90     }
91 }
```

程序运行结果如图 3.15 所示。

(a) 打开进度对话框　　　　　　　　(b) 打开时间对话框

图 3.15　进度对话框和时间对话框示例

3.4 Fragment

■ 3.4.1 动态加载 Fragment 对象

Android 在 3.0 版本之后引入了 Fragment（片段）的概念，其主要目的是为了适配不同设备的大屏幕、支持组件的动态加载和更灵活的 UI 设计。

Fragment 对象在应用中一般不能独立存在，它是页面 Activity 的一部分，必须嵌入在某一个 Activity 页面之中。

Fragment 嵌入到 Activity 页面中分为静态加载和动态加载两种形式。静态加载是在设计 Activity 页面时，把 Fragment 组件设计到界面之中；动态加载则是再设计一个占位组件，用于给动态加载的 Fragment 组件占据位置，在运行程序时，根据需要来决定是否加载 Fragment 组件。

实现动态加载 Fragment 组件的基本方法如下。

（1）创建一个 Fragment 管理器对象 FragmentManager。

```
FragmentManager fragmentManager = getSupportFragmentManager();
```

（2）创建一个 FragmentTransaction 事务处理对象。

```
fragmentTransaction = fragmentManager.beginTransaction();
```

（3）用 replace( )方法指定加载的 Fragment 组件。

```
fragmentTransaction.replace(R.id.fragment_id, new FirstFragment());
```

其中，fragment_id 是 Fragment 组件的 id，FirstFragment 是 Fragment 组件的类名。

（4）把 Fragment 组件加载到堆栈。

```
fragmentTransaction.addToBackStack(null);
```

（5）提交事务，执行动态加载。

```
fragmentTransaction.commit();
```

【例 3-7】 设计一个动态加载 Fragment 的示例。

（1）新建应用程序，系统自动生成应用程序框架。

（2）新建两个 Fragment。

右击应用程序的 APP 选项，选择 New→Fragment→Fragment(Blank)选项，如图 3.16 所示。新建的两个 Fragment 分别命名为 FirstFragment 和 SecondFragment。

在 Fragment 的界面布局文件中，设置 Fragment 的 background 属性值，设置其背景颜色。修改界面布局中 TextView 的 text 属性值，分别设为 "这是第 1 个 Fragment" 和 "这是第 2 个 Fragment"。

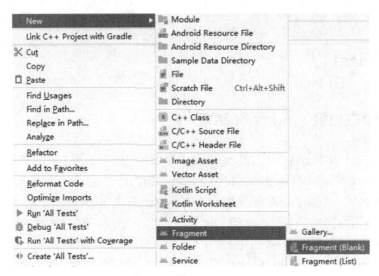

图 3.16　新建 Fragment

（3）设计 Activity 页面的界面布局。

Activity 页面的界面布局设计如图 3.17 所示。设置一个空的水平线性布局 LinearLayout（horizontal），其 id 设为 fragment_id，这是一个占位组件，用于给动态加载的 Fragment 组件占据位置。

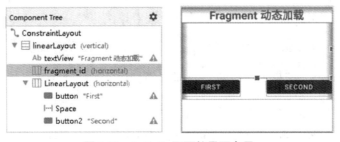

图 3.17　Activity 页面的界面布局

（4）设计主控制程序。

```
1   package com.example.ex3_7_fragment;
2   import androidx.appcompat.app.AppCompatActivity;
3   import android.os.Bundle;
4   import androidx.fragment.app.Fragment;
5   import androidx.fragment.app.FragmentManager;
6   import androidx.fragment.app.FragmentTransaction;
7   import android.view.View;
8   import android.widget.Button;

9   public class MainActivity extends AppCompatActivity {
10      Button buttonOne,buttonTwo;
11      FragmentManager fragmentManager = null;    //Fragment 管理器
12      FragmentTransaction fragmentTransaction;   //Fragment 事务处理
13      @Override
```

```
14    protected void onCreate(Bundle savedInstanceState) {
15        super.onCreate(savedInstanceState);
16        setContentView(R.layout.activity_main);
17        buttonOne = (Button) findViewById(R.id.button);
18        buttonTwo = (Button) findViewById(R.id.button2);
19        buttonOne.setOnClickListener(new mClick());
20        buttonTwo.setOnClickListener(new mClick());
21    }

22    private void initFragment() {
23        Fragment fragment = new Fragment(); //实例化 Fragment
24        fragmentManager = getSupportFragmentManager();//得到 Fragment 管理器对象
25        //开始 fragmnet 的事务处理
26        fragmentTransaction = fragmentManager.beginTransaction();
27        fragmentTransaction.add(R.id.fragment_id, fragment);
          //fragment_id 是布局中给 Fragment 占位置的组件
28    }

29    class mClick implements View.OnClickListener{
30        public void onClick(View v) {
31            initFragment();
32            if(v==buttonOne) {
33                //加载第一个 Fragment
34                fragmentTransaction.replace(R.id.fragment_id, new FirstFragment());
35                //把 FirstFragment 放到堆栈里
36                fragmentTransaction.addToBackStack(null);
37            }
38            if(v == buttonTwo) {
39                //加载第二个 Fragment
40                fragmentTransaction.replace(R.id.fragment_id, new SecondFragment());
41                //把 SecondFragment 放到堆栈里
42                fragmentTransaction.addToBackStack(null);
43            }
44            fragmentTransaction.commit();//提交事务
45        }
46    }
47 }
```

程序运行结果如图 3.18 所示。

图 3.18 动态加载 Fragment

3.4.2 底部导航栏

在实际编写应用程序时，不必每一个 Activity 都要从零开始，利用好系统自带的模板往往可以起到事半功倍的效果。下面介绍使用 Bottom Navigation Activity 模板来完成简单的底部导航栏功能。

【例 3-8】 设计一个底部导航的简单应用程序。

1）系统自动生成底部导航页面

在创建应用程序时，选择 Bottom Navigation Activity 选项，如图 3.19 所示。

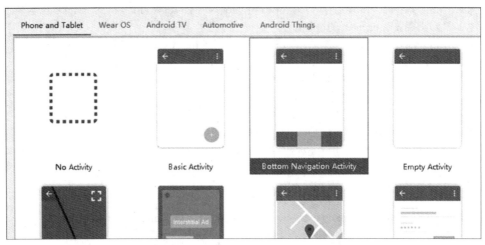

图 3.19 选择 Bottom Navigation Activity 选项

2）生成应用程序的文件目录结构

（1）生成 Java 目录下的 Fragment 文件。

在应用程序的 java\<包名>\ui 目录下，生成了 DashboardFragment、HomeFragment 和 NotificationsFragment 等 3 个 Fragment 文件，如图 3.20 所示。

图 3.20 生成的 3 个 Fragment 文件

（2）生成 res 资源的文件目录结构。

在应用程序的界面布局目录 res\layout 下，除了 activity_main.xml 外，还生成了 3 个

Fragment 的布局文件。

在菜单目录 res\menu 下，生成了页面底部菜单文件 bottom_nav_menu.xml。

在导航目录 res\navigation 下，生成了包含 3 个 Fragment 组件的导航布局文件 mobile_navigation.xml。

res 资源的文件目录结构如图 3.21 所示。

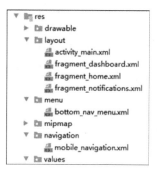

图 3.21　res 资源的文件目录结构

3）新增一个 Fragment

（1）在 java\ui 目录下，右击 ui 目录项，选择"新建"→package 项。在 New Package 对话框中，填写新的包名为 com.example.ex3_8.news。这时，在 Android Studio 系统中可以看到，在 ui 目录下新建了 news 目录。

（2）右击 ui\news 目录，在弹出的菜单中选择 New→Fragment→fragment(with ViewModel) 选项，在 New Android Component 对话框中，填写 Fragment Name 为 NewsFragment，如图 3.22 所示。

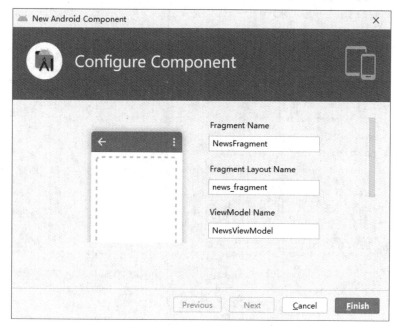

图 3.22　新建 NewsFragment

4）修改生成的模板文件

（1）修改 string.xml，增加 news 的底部导航标签文字内容。

打开 res\values\strings.xml 文件，增加一个<string name="title_news">标签，其代码如下：

```
1    <resources>
2        <string name="app_name"> ex3_8</string>
3        <string name="title_home">首页</string>
4        <string name="title_dashboard">图片</string>
5        <string name="title_notifications">通知</string>
6        <string name="title_news">新闻</string>    ← 新增"新闻"项
7    </resources>
```

（2）修改导航布局文件，增加 Fragment 到导航界面中。

打开 res\navigation\mobile_navigation.xml 文件，新增 NewsFragment 到导航界面中。

```
1    <fragment
2        android:id="@+id/navigation_news"
3        android:name="com.example.ex3_8_fragment.ui.news.NewsFragment"
4        android:label="@string/title_news"
5        tools:layout="@layout/news_fragment" />
```

完整的 mobile_navigation.xml 代码如下。

```
1    <?xml version="1.0" encoding="utf-8"?>
2    <navigation xmlns:android="http://schemas.android.com/apk/res/android"
3        xmlns:app="http://schemas.android.com/apk/res-auto"
4        xmlns:tools="http://schemas.android.com/tools"
5        android:id="@+id/mobile_navigation"
6        app:startDestination="@+id/navigation_home">
7        <fragment
8            android:id="@+id/navigation_home"
9            android:name="com.example.ex3_8_fragment.ui.home.HomeFragment"
10           android:label="@string/title_home"
11           tools:layout="@layout/fragment_home" />
12       <fragment
13           android:id="@+id/navigation_dashboard"
14           android:name="com.example.ex3_8_fragment.ui.dashboard.
             DashboardFragment"
15           android:label="@string/title_dashboard"
16           tools:layout="@layout/fragment_dashboard" />
17       <fragment
18           android:id="@+id/navigation_notifications"
19           android:name="com.example.ex3_8_fragment.ui.notifications.
             NotificationsFragment"
20           android:label="@string/title_notifications"
21           tools:layout="@layout/fragment_notifications" />
22       <fragment
23         android:id="@+id/navigation_news"
```

```
24         android:name="com.example.ex3_8_fragment.ui.news.NewsFragment"
25         android:label="@string/title_news"
26         tools:layout="@layout/news_fragment" />
27 </navigation>
```

内容扩展

（3）修改菜单项。

打开菜单文件 res\menu\bottom_nav_menu.xml，增加一个<item>菜单项。

```
1  <item
2      android:id="@+id/navigation_news"
3      android:icon="@drawable/abc_vector_test"
4      android:title="@string/title_news"  />
```

完整的 bottom_nav_menu.xml 代码如下。

```
1  <?xml version="1.0" encoding="utf-8"?>
2  <menu xmlns:android="http://schemas.android.com/apk/res/android">
3      <item
4          android:id="@+id/navigation_home"
5          android:icon="@drawable/ic_home_black_24dp"
6          android:title="@string/title_home" />
7      <item
8          android:id="@+id/navigation_dashboard"
9          android:icon="@drawable/ic_dashboard_black_24dp"
10         android:title="@string/title_dashboard" />
11     <item
12         android:id="@+id/navigation_notifications"
13         android:icon="@drawable/ic_notifications_black_24dp"
14         android:title="@string/title_notifications" />
15     <item
16         android:id="@+id/navigation_news"
17         android:icon="@drawable/abc_vector_test"
18         android:title="@string/title_news"  />
19 </menu>
```

该菜单文件的设计界面新增了一个"新闻"菜单项，如图 3.23 所示。

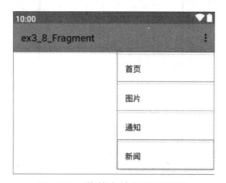

图 3.23　菜单文件的设计界面

（4）修改主控制程序 MainActivity.java。

```
1   package com.example.ex3_8;
2   import androidx.appcompat.app.AppCompatActivity;
3   import android.os.Bundle;
4   import com.google.android.material.bottomnavigation.BottomNavigationView;
5   import androidx.navigation.NavController;
6   import androidx.navigation.Navigation;
7   import androidx.navigation.ui.AppBarConfiguration;
8   import androidx.navigation.ui.NavigationUI;

9   public class MainActivity extends AppCompatActivity {
10      @Override
11      protected void onCreate(Bundle savedInstanceState) {
12         super.onCreate(savedInstanceState);
13         setContentView(R.layout.activity_main);
14         BottomNavigationView navView = findViewById(R.id.nav_view);
15         AppBarConfiguration appBarConfiguration = new AppBarConfiguration.Builder(
16                 R.id.navigation_home,
17                 R.id.navigation_dashboard,
18                 R.id.navigation_notifications,
19                 R.id.navigation_news    ← 新增"新闻"Fragment 项
20         ).build();
21         NavController navController = Navigation.findNavController(this,
22                 R.id.nav_host_fragment);
23         NavigationUI.setupActionBarWithNavController(this, navController,
24                 appBarConfiguration);
25         NavigationUI.setupWithNavController(navView, navController);
26      }
27   }
```

运行程序可以看到，底部导航栏中新增了一个"新闻"选项，如图 3.24 所示。

图 3.24　底部导航栏中新增的"新闻"选项

习题 3

1. 设计一个具有两个页面的程序,第 1 个页面显示一张封面的图片,第 2 个页面显示"欢迎进入本系统",这两个页面之间能相互切换。
2. 设计一个具有 3 个选项的菜单程序,当点击每个选项时,分别跳转到 3 个不同的页面。
3. 设计一个具有计算器功能的对话框程序。

第4章 图形与多媒体处理

4.1 绘制几何图形

4.1.1 几何图形绘制类

在 Android 系统中，绘制几何图形需要用到一些绘图工具，这些绘图工具都在 android.graphics 包中。下面介绍这些绘图工具类的常用方法和属性。

1. 画布类 Canvas

画布类 Canvas 是 Android 绘制几何图形的主要工具，其常用方法如表 4-1 所示。

表 4-1 画布类 Canvas 的常用方法

方法	功能
Canvas()	创建一个空的画布，可以使用 setBitmap()方法来设置绘制具体的画布
Canvas(Bitmap bitmap)	以 bitmap 为对象创建一个画布，将内容都绘制在 bitmap 上，bitmap 不得为 null
drawColor()	设置 Canvas 的背景颜色
setBitmap()	设置具体画布
clipRect()	设置显示区域，即设置裁剪区
rotate()	旋转画布
skew()	设置偏移量
drawLine(float x1, float y1, float x2, float y2)	从点(x1, y1)到点(x2, y2)的直线
drawCircle(float x, float y, float radius, Paint paint)	以(x, y)为圆心、radius 为半径画圆
drawRect(float x1, float y1, float x2, float y2, Paint paint)	从左上角(x1, y1)到右下角(x2, y2)的矩形
drawText(String text, float x, float y, Paint paint)	写文字
drawBitmap (Bitmap bitmap, float x, float y, Paint paint)	以左上角坐标（x, y）为顶点绘制 Bitmap 图像
drawPath(Path path, Paint paint)	从一点到另一点的连接路径线段

2. 画笔类 Paint

画笔类 Paint 用来描述所绘制图形的颜色和风格，如线条宽度、颜色等信息，其常用的方法如表 4-2 所示。

表 4-2 画笔类 Paint 的常用方法

方　法	功　能
Paint()	构造方法，创建一个辅助画笔对象
setColor(int color)	设置颜色
setStrokeWidth(float width)	设置画笔宽度
setTextSize(float textSize)	设置文字尺寸
setAlpha(int a)	设置透明度 Alpha 值
setAntiAlias(boolean b)	除去边缘锯齿，取 true 值
paint.setStyle(Paint.Style style)	设置图形为空心(Paint.Style.STROKE)或实心(Paint.Style.FILL)

3. 点到点的连线路径 Path

在绘制由一些线段组成的图形（如三角形、四边形等）时，需要用 Path 类来描述线段路径，其常用方法如表 4-3 所示。

表 4-3 连线路径 Path 的常用方法

方　法	功　能
lineTo(float x, float y)	从当前点到指定点画连线
moveTo(float x, float y)	移动到指定点
close()	关闭绘制连线路径

■ 4.1.2 几何图形绘制过程

在 Android 中绘制几何图形的一般过程如下：
（1）创建一个 View 的子类，并重写 View 类的 onDraw()方法。
（2）在 View 的子类视图中使用画布对象 Canvas 绘制各种图形。
（3）使用 invalidate()方法刷新画面。

【例 4-1】 绘制几何图形示例。

本例继承自 Android.view.View 的 TestView 类，重写 View 类的 onDraw()方法，在 onDraw() 方法中运用 Paint 对象（画笔）的不同设置值，在 Cavas（画布）上分别绘制矩形、圆形、三角形和文字。

其源程序如下：

```
1    package com.example.ex4_1;
2    import androidx.appcompat.app.AppCompatActivity;
3    import android.os.Bundle;
4    import android.content.Context;
```

```
5    import android.graphics.Canvas;
6    import android.graphics.Color;
7    import android.graphics.Paint;
8    import android.graphics.Path;
9    import android.view.View;

10   public class MainActivity extends AppCompatActivity
11   {
12       @Override
13       public void onCreate(Bundle savedInstanceState)
14       {
15           super.onCreate(savedInstanceState);
16           setContentView(R.layout.activity_main);
17           TestView tView = new TestView(this);
18           setContentView(tView);     //显示自定义的组件
19   }
20   //下面的代码为创建一个可视的组件（View 的子类）
21   private class TestView extends View
22   {
23       public TestView(Context context)
24       {
25           super(context);   //调用父类的构造方法
26       }
27       /*重写 onDraw（）*/
28       @Override
29       protected void onDraw(Canvas canvas)
30       {
31       canvas.drawColor(Color.CYAN);           //设置画布的背景颜色为青色
32       Paint paint = new Paint();              //定义画笔
33       paint.setStrokeWidth(3);                //设置画笔的宽度（线条的宽度）
34       /* 画空心矩形  */
35       paint.setStyle(Paint.Style.STROKE);     //设置绘制的图形为空心
36       canvas.drawRect(10, 10, 70, 70, paint);
37       /* 画实心矩形  */
38       paint.setStyle(Paint.Style.FILL);       ///设置绘制的图形为实心
39       canvas.drawRect(100,10,170,70,paint);
40       /* 画实心圆  */
41       paint.setColor(Color.BLUE);             //设置画笔颜色为蓝色
42       paint.setAntiAlias(true);               //去锯齿
43       canvas.drawCircle(100,120,30,paint);
                                                 //画圆心为（100,120）、半径为 30 的实心圆
44       paint.setColor(Color.WHITE);            //设置画笔为白色
45       canvas.drawCircle(91,111,6,paint);      //画出半径为 6 的小圆点
46       /* 画三角形 */
47       paint.setColor(Color.RED);              //设置画笔为红色
48       Path path=new Path();
49       path.moveTo(100, 170);     //使用 Path 时，第 1 个点都是用 move(x, y)
```

```
50      path.lineTo(70, 230);      //以后的点均用lineTo(x, y),即向点(x,y)连线
51      path.lineTo(130,230);
52      path.close();                              //围成封闭区域
53      canvas.drawPath(path,paint);               //用Canvas画出多边形
54      /* 文字 */
55      paint.setTextSize(28);
56      paint.setColor(Color.BLUE);
57      canvas.drawText(getResources().getString(R.string.hello_world),
58                 30,270,paint);
59    }
60  }
61 }
```

程序运行结果如图 4.1 所示。

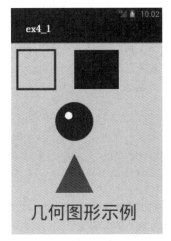

图 4.1 绘制几何图形示例

4.1.3 自定义组件

视频讲解

在 Android 中，可以通过 View 类的子类自定义组件，然后再添加到布局界面中。下面通过示例详细说明设计方法。

【例 4-2】自定义一个组件，再通过布局界面显示出来。

主要设计步骤如下：

（1）编写 View 的子类 TestView。

（2）把 TestView 添加到布局界面文件中。

其添加的格式为：

<包名.自定义的组件名称>

例如：

(3) 在主程序 MainActivity.java 中建立 TestView 对象与布局文件的关联。

具体设计如下：

(1) 在应用程序的 com.example.ex4_2 目录下，新建 Java 文件，编写 View 的子类 TestView。

```java
1   package com.example.ex4_2;
2   import android.graphics.Canvas;
3   import android.graphics.Color;
4   import android.graphics.Paint;
5   import android.content.Context;
6   import android.util.AttributeSet;
7   import android.view.View;
8
9   class TestView extends View
10  {
11      public TestView(Context context, AttributeSet attrs)
12      {   super(context, attrs);   }
13      /* 重写View的抽象方法 onDraw()方法 */
14      protected void onDraw(Canvas canvas)
15      {
16          canvas.drawColor(Color.CYAN);         //设置组件的背景颜色为青色
17          Paint paint=new Paint();              //定义画笔
18          paint.setStyle(Paint.Style.FILL);     //设置画实心图形
19          paint.setAntiAlias(true);             //去锯齿
20          paint.setColor(Color.BLUE);           //设置画笔颜色为蓝色
21          canvas.drawCircle(100,120,30,paint);  //画圆心为（100,120）、半径为30的实心圆
22          paint.setColor(Color.WHITE);
23          canvas.drawCircle(91,111,6,paint);    //在实心圆上画一个小白点
24      }
25  }
```

(2) 在表现层布局文件 activity_main.xml 中，添加所设计的组件 TestView 类。

```xml
1   <?xml version="1.0" encoding="utf-8"?>
2   <androidx.constraintlayout.widget.ConstraintLayout
3       xmlns:android="http://schemas.android.com/apk/res/android"
4       xmlns:app="http://schemas.android.com/apk/res-auto"
5       xmlns:tools="http://schemas.android.com/tools"
6       android:layout_width="match_parent"
7       android:layout_height="match_parent"
8       tools:context=".MainActivity">
9
10      <com.example.ex4_2.TestView
11          android:id="@+id/testview1"
12          android:layout_width="wrap_content"
13          android:layout_height="wrap_content"
14      />
```

←添加自定义组件

```
14    </androidx.constraintlayout.widget.ConstraintLayout>
```

这时，点击菜单 Build 中的 Rebuild Project 选项，重新编译应用程序，可以在编辑器的预览中看到所自定义的组件，同时，在工具面板中可以看到 Project 选项，该选项下有自定义的 TestView 组件，如图 4.2 所示。

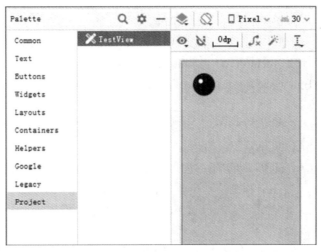

图 4.2　在编辑器中预览自定义的组件

（3）在主程序 MainActivity.java 中建立 TestView 对象与布局文件的关联。

```
1     package com.example.ex4_2;
2     import androidx.appcompat.app.AppCompatActivity;
3     import android.os.Bundle;
4     public class MainActivity extends Activity {
5         TestView tView = null;
6     @Override
7     public void onCreate(Bundle savedInstanceState) {
8         super.onCreate(savedInstanceState);
9         setContentView(R.layout.activity_main);
10        tView =(TestView)findViewById(R.id.testview1);
11    }
12    }
```

运行程序，其运行结果与编辑器预览中的预览结果一致。

4.2　触摸屏事件处理

智能移动设备的触摸屏事件（在模拟器中为鼠标事件）分为简单的触摸屏事件和手势识别。下面分别介绍这些事件的处理方法。

4.2.1 简单的触摸屏事件

简单的触摸屏事件是指触摸屏按下、抬起和移动事件（在模拟器中为鼠标事件）。在 Android 系统中，通过 OnTouchListener 监听接口来处理屏幕事件，当在 View 的范围内触摸按下、抬起或滑动等动作时都会触发该事件。

在设计简单触摸屏事件程序时，要实现 android.view.View.OnTouchListener 接口，并重写该接口的监听方法 onTouch(View v, MotionEvent event)。

在监听方法 onTouch(View v, MotionEvent event)中，参数 v 为事件源对象，参数 event 为事件对象，事件对象为下列常数之一。

- MotionEvent.ACTION_DOWN：在屏幕上点击。
- MotionEvent.ACTION_UP：抬起。
- MotionEvent.ACTION_MOVE：在屏幕上滑动。

【例 4-3】 设计一个在屏幕上移动小球的程序。

设计一个自定义组件继承于 Android.view.View 的图形绘制类 TestView，在该视图组件中绘制一个小球。

（1）设计图形绘制类 TestView（自定义组件）。

在该组件中绘制小球图形，再增加一个获取手指在屏幕上位置的传值函数 getXY()，其代码如下。

```
1   package com.example.ex4_3;
2   import   (略)
3   class TestView extends View
4   {
5       int x=150,y=50;              ◁── 定义小球初始坐标
6       public TestView(Context context, AttributeSet attrs)
7       {  super(context, attrs);  }
8       void getXY(int _x, int _y)
9       {
10          x = _x;                  ◁── 由触摸屏事件传递小球坐标位置
11          y = _y;
12      }
13      @Override
14      protected void onDraw(Canvas canvas)
15      {
16          super.onDraw(canvas);
17          canvas.drawColor(Color.CYAN);   /*设置背景为青色*/
18          Paint paint=new Paint();
19          paint.setAntiAlias(true);       /*去锯齿*/
20          paint.setColor(Color.BLACK);    /*设置 paint 的颜色*/
21          canvas.drawCircle(x, y, 30, paint);  /*画一个实心圆*/
22          paint.setColor(Color.WHITE);    /*画实心圆上的小白点*/
23          canvas.drawCircle(x-9, y-9, 6, paint);
```

```
24      }
25  }
```

（2）把自定义组件添加到布局文件 activity_main.xml 中。

```xml
1   <?xml version="1.0" encoding="utf-8"?>
2   <androidx.constraintlayout.widget.ConstraintLayout
3       xmlns:android="http://schemas.android.com/apk/res/android"
4       xmlns:app="http://schemas.android.com/apk/res-auto"
5       xmlns:tools="http://schemas.android.com/tools"
6       android:layout_width="match_parent"
7       android:layout_height="match_parent"
8       tools:context=".MainActivity">

9       <com.example.ex4_3.TestView
10          android:id="@+id/testview1"
11          android:layout_width="wrap_content"
12          android:layout_height="wrap_content"
13          />
14  </androidx.constraintlayout.widget.ConstraintLayout>
```

（9~13 行：添加自定义组件）

（3）设计主程序 MainActivity.java。

在主程序中，设计一个实现监听触摸屏事件的方法 onTouch(View v, MotionEvent event)，该方法监听并获取触摸屏幕的坐标位置，并把坐标值传递给图形绘制类 TestView，由 TestView 在该位置重绘小球。

```java
1   package com.example.ex4_3;
2   import androidx.appcompat.app.AppCompatActivity;
3   import android.os.Bundle;
4   import android.view.MotionEvent;
5   import android.view.View;
6   public class MainActivity extends Activity {
7       TestView tView = null;
8       @Override
9       public void onCreate(Bundle savedInstanceState) {
10          super.onCreate(savedInstanceState);
11          setContentView(R.layout.activity_main);
12          tView =(TestView)findViewById(R.id.testview1);
13          tView.setOnTouchListener(new mOnTouch());
14      }

15      private class mOnTouch implements OnTouchListener    ← 定义触摸屏事件
16      {
17          public boolean onTouch(View v, MotionEvent event)
18          {
```

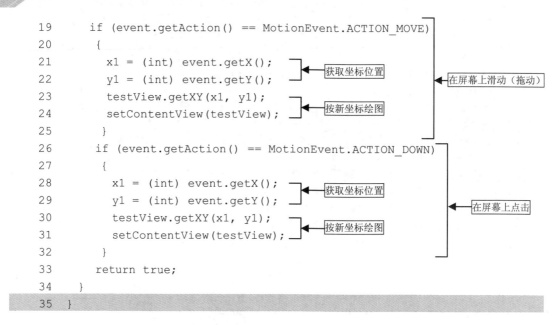

```
19      if (event.getAction() == MotionEvent.ACTION_MOVE)
20      {
21        x1 = (int) event.getX();
22        y1 = (int) event.getY();
23        testView.getXY(x1, y1);
24        setContentView(testView);
25      }
26      if (event.getAction() == MotionEvent.ACTION_DOWN)
27      {
28        x1 = (int) event.getX();
29        y1 = (int) event.getY();
30        testView.getXY(x1, y1);
31        setContentView(testView);
32      }
33      return true;
34    }
35  }
```

程序运行结果如图 4.3 所示，手指（或鼠标）在屏幕上滑动时，小球将随手指（或鼠标）移动。当点击屏幕时，小球将被移动到所点击的位置。

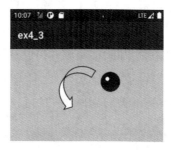

图 4.3 手指（或鼠标）在屏幕上滑动时，小球随之移动

4.2.2 手势识别

所谓手势识别，就是识别手指（或鼠标）在屏幕上滑动时的轨迹。下面介绍应用触摸屏技术实现手势识别的方法。

视频讲解

【例 4-4】 应用触摸屏技术设计一个具有左右滑屏切换页面功能的程序。

根据触摸屏事件，当手指在触摸屏上按下时获得一个坐标点：

```
x1 = event.getX();
```

手指在屏幕上滑动，当抬起手指时，获得另一个坐标点：

```
X2 = event.getX();
```

手指向左滑动屏幕时，x1>x2；手指向右滑动屏幕时，x2>x1。

由于屏幕像素的距离太小，为了避免手指在屏幕上抖动而触发屏幕事件，通常使用 x1 − x2 > 50 及 x2 − x1 > 50 来判断手指滑动的方向。

程序设计过程如下。
(1) 新建两个页面，页面的界面布局略。
(2) 设计第 1 个页面的控制程序 MainActivity.java。

```
1   package com.example.ex4_4;
2   import androidx.appcompat.app.AppCompatActivity;
3   import android.os.Bundle;
4   import android.view.MotionEvent;
5   import android.content.Intent;

6   public class MainActivity extends AppCompatActivity {
7       float x1, x2;                    //仅作水平滑动屏幕,故只需设 x 轴坐标变量
8       Intent int1;
9       @Override
10      protected void onCreate(Bundle savedInstanceState) {
11          super.onCreate(savedInstanceState);
12          setContentView(R.layout.activity_main);
13      }
14      public boolean onTouchEvent(MotionEvent event) {
15          if(event.getAction() == MotionEvent.ACTION_DOWN){  //按下手指
16              x1 = event.getX();        //获取第 1 个坐标值
17          }
18          if(event.getAction() == MotionEvent.ACTION_UP) {   //抬起手指
19              x2 = event.getX();        //获取第 2 个坐标值
20              int1 = new Intent(MainActivity.this,MainActivity2.class);
21              if(x1 - x2 > 50) {        //向左滑屏
22                  startActivity(int1);
23              }
24          }
25          return super.onTouchEvent(event);
26      }
27  }
```

(3) 设计第 2 个页面的控制程序 MainActivity2.java。

```
1   package com.example.ex4_4;
2   import androidx.appcompat.app.AppCompatActivity;
3   import android.os.Bundle;
4   import android.view.MotionEvent;
5   import android.content.Intent;

6   public class MainActivity2 extends AppCompatActivity {
7       float x1, x2;
8       Intent int2;
9       @Override
10      protected void onCreate(Bundle savedInstanceState) {
11          super.onCreate(savedInstanceState);
```

```
12          setContentView(R.layout.activity_main2);
13      }
14      public boolean onTouchEvent(MotionEvent event) {
15        if(event.getAction() == MotionEvent.ACTION_DOWN) {    //按下手指
16          x1 = event.getX();         //获取第 1 个坐标值
17        }
18        if(event.getAction() == MotionEvent.ACTION_UP) {      //抬起手指
19          x2 = event.getX();         //获取第 2 个坐标值
20          int1 = new Intent(MainActivity2.this,MainActivity.class);
21          if(x2 - x1 > 50) {         //向右滑屏
22            startActivity(int2);
23          }
24        }
25        return super.onTouchEvent(event);
26      }
27    }
```

程序运行结果如图 4.4 所示。

图 4.4　手指在屏幕上滑动时切换页面

视频讲解

【例 4-5】　设计一个简单的画图板程序。

将事先准备好的一个空白画面的图片文件 xbh.jpg 保存到应用项目的 res\drawable 目录下，作画图的空白背景之用。

（1）设计布局文件。

```
1   <?xml version="1.0" encoding="utf-8"?>
2   <androidx.constraintlayout.widget.ConstraintLayout
3       xmlns:android="http://schemas.android.com/apk/res/android"
4       xmlns:app="http://schemas.android.com/apk/res-auto"
5       xmlns:tools="http://schemas.android.com/tools"
6       android:layout_width="match_parent"
7       android:layout_height="match_parent"
8       tools:context=".MainActivity">
9       <com.example.ex4_2.TestView
10          android:id="@+id/testview1"
11          android:layout_width="wrap_content"
12          android:layout_height="wrap_content"
```

导入自定义 View，注意要带包名

```
13        />
14        <Button
15            android:id="@+id/button"
16            android:layout_width="wrap_content"
17            android:layout_height="wrap_content"
18            android:layout_marginStart="128dp"
19            android:layout_marginLeft="128dp"
20            android:layout_marginTop="16dp"
21            android:text="清屏"
22            app:layout_constraintStart_toStartOf="parent"
23            app:layout_constraintTop_toBottomOf="@+id/handWrite" />
24    </androidx.constraintlayout.widget.ConstraintLayout>
```

(2) 设计主控文件 MainActivity.java。

```
1   package com.example.ex4_5;
2   import androidx.appcompat.app.AppCompatActivity;
3   import android.os.Bundle;
4   import android.view.View;
5   import android.view.View.OnClickListener;
6   import android.widget.Button;
7   public class MainActivity extends AppCompatActivity
8   {
9       private HandWrite handWrite = null;
10      private Button clear = null;
11      @Override
12      public void onCreate(Bundle savedInstanceState)
13      {
14          super.onCreate(savedInstanceState);
15          setContentView(R.layout.main);
16          handWrite = (HandWrite)findViewById(R.id.handwriteview);   ← 关联View组件
17          clear = (Button)findViewById(R.id.clear);
18          clear.setOnClickListener(new mClick());
19      }
20      private class mClick implements OnClickListener
21      {
22          public void onClick(View v)
23          {
24              handWrite.clear();   ← 清屏
25          }
26      }
27  }
```

(3) 设计自定义组件 HandWrite.java。
该组件用于记录在屏幕上滑动的轨迹，实现在画图板上涂鸦的功能。

```
1   package com.example.ex4_5;
2   import android.content.Context;
```

```
3     import android.graphics.Bitmap;
4     import android.graphics.BitmapFactory;
5     import android.graphics.Canvas;
6     import android.graphics.Color;
7     import android.graphics.Paint;
8     import android.graphics.Paint.Style;
9     import android.util.AttributeSet;
10    import android.view.MotionEvent;
11    import android.view.View;
12    public class HandWrite extends View        ←自定义 View 组件 HandWrite
13    {
14        Paint paint = null;                    //定义画笔
15        Bitmap originalBitmap = null;          //存放原始图像
16        Bitmap new1_Bitmap = null;             //存放从原始图像复制的位图图像
17        Bitmap new2_Bitmap = null;             //存放处理后的图像
18        float startX = 0, startY = 0;          //画线的起点坐标
19        float clickX = 0, clickY = 0;          //画线的终点坐标
20        boolean isMove = true;                 //设置是否画线的标记
21        boolean isClear = false;               //设置是否清除涂鸦的标记
22        int color = Color.GREEN;               //设置画笔的颜色（绿色）
23        float strokeWidth = 5.0f;              //设置画笔的宽度
24      public HandWrite(Context context, AttributeSet attrs)
25      {
26         super(context, attrs);
27         originalBitmap = BitmapFactory      ←从资源中获取原始图像
28              .decodeResource(getResources(), R.drawable.xbh);
29         originalBitmap = originalBitmap
30              .copy(Bitmap.Config.ARGB_8888, true);
31         new1_Bitmap = Bitmap.createBitmap(originalBitmap);  ←建立原始图像的位图
32      }
33       public void clear(){
34         isClear = true;
35         new2_Bitmap = Bitmap.createBitmap(originalBitmap);
36         invalidate();    //刷新界面                                ←清除涂鸦
37       }
38       public void setstyle(float strokeWidth){
39           this.strokeWidth = strokeWidth;
40        }
41       @Override
42       protected void onDraw(Canvas canvas)
43       {
44         super.onDraw(canvas);                                       ←显示绘图
45         canvas.drawBitmap(HandWriting(new1_Bitmap), 0, 0,null);
46       }
47       public Bitmap HandWriting(Bitmap o_Bitmap)     ←记录绘制图形
48       {
49         Canvas canvas = null;   ←定义画布
```

```
50      if(isClear)
51      {
52          canvas = new Canvas(new2_Bitmap);    // 创建绘制新图形的画布
53      }
54      else{
55          canvas = new Canvas(o_Bitmap);        // 创建绘制原图形的画布
56      }
57      paint = new Paint();
58      paint.setStyle(Style.STROKE);
59      paint.setAntiAlias(true);                  // 定义画笔
60      paint.setColor(color);
61      paint.setStrokeWidth(strokeWidth);
62      if(isMove)
63      {
64          canvas.drawLine(startX, startY, clickX, clickY, paint);    // 在画布上画线条
65      }
66      startX = clickX;
67      startY = clickY;
68      if(isClear)
69      {
70          return new2_Bitmap;                    // 返回新绘制的图像
71      }
72      return o_Bitmap;                           // 若清屏，则返回原图像
73  }
74  @Override
75  public boolean onTouchEvent(MotionEvent event)    // 定义触摸屏事件
76  {
77      clickX = event.getX();
78      clickY = event.getY();                     // 获取触摸坐标位置
79      if(event.getAction() == MotionEvent.ACTION_DOWN)    // 按下屏幕时无绘图
80      {
81          isMove = false;
82          invalidate();
83          return true;
84      }
85      else if(event.getAction() == MotionEvent.ACTION_MOVE)
86      {
87          isMove = true;
88          invalidate();                          // 记录在屏幕上滑动的轨迹
89          return true;
90      }
91      return super.onTouchEvent(event);
92  }
93 }
```

程序运行结果如图 4.5 所示，当手指（或鼠标）在屏幕上滑动时，记录下滑动的轨迹，在图片上涂鸦。点击"清屏"按钮，则清除图片上的痕迹。

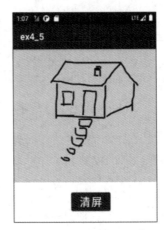

图 4.5 在画图板上涂鸦

4.3 音频播放

4.3.1 多媒体处理包

Android 系统提供了针对常见多媒体格式的 API，可以非常方便地操作图片、音频、视频等多媒体文件，也可以操作 Android 终端的录音、摄像设备。这些多媒体处理 API 均位于 android.media 包中。android.media 包中的主要类如表 4-4 所示。

表 4-4 android.media 包中的主要类

类名或接口名	说明
MediaPlayer	支持流媒体，用于播放音频和视频
MediaRecorder	用于录制音频和视频
Ringtone	用于播放可用作铃声和提示音的短声音片段
AudioManager	负责控制音量
AudioRecord	用于记录从音频输入设备产生的数据
JetPlayer	用于存储 JET 内容的回放和控制
RingtoneManager	用于访问响铃、通知和其他类型的声音
SoundPool	用于管理和播放应用程序的音频资源

4.3.2 多媒体处理播放器

1. MediaPlayer 类的常用方法

多媒体处理播放器类（MediaPlayer）是 Android 系统多媒体 android.media 包中的类，主要用于控制音频文件、视频文件或流媒体的播放。MediaPlayer 类的常用方法如表 4-5 所示。

第4章 图形与多媒体处理

表 4-5 MediaPlayer 类的常用方法

方　　法	说　　明
create( )	创建多媒体播放器
getCurrentPosition()	获得当前播放位置
getDuration()	获得播放文件的时间
getVideoHeight()	播放视频高度
getVideoWidth()	播放视频宽度
isLooping()	是否循环播放
isPlaying()	是否正在播放
Pause()	暂停
Prepare()	准备播放文件，进行同步处理
prepareAsync()	准备播放文件，进行异步处理
release()	释放 MediaPlayer 对象
reset()	重置 MediaPlayer 对象
seekTo()	指定播放文件的播放位置
setDataSource()	设置多媒体数据来源
setVolume()	设置音量
setOnCompletionListener()	监听播放文件播放完毕
start()	开始播放
stop()	停止播放

2. MediaPlayer 对象的生命周期

通常把一个对象从创建、使用直到释放该对象的过程称为该对象的生命周期，把 MediaPlayer 对象的创建、初始化、同步处理、开始播放、播放结束的运行过程称为 MediaPlayer 的生命周期。MediaPlayer 对象的生命周期如图 4.6 所示。

从图 4.6 可以看出，当一个 MediaPlayer 对象刚创建或调用了 reset()方法后，它处于 Idle（空闲）状态；当调用了 release()方法后，它处于释放（结束）状态。这两种状态之间是 MediaPlayer 对象的生命周期。一个 MediaPlayer 对象在处于空闲状态时是不能进行播放操作的，还必须经过初始化、同步阶段才能进行播放操作。

4.3.3 播放音频文件

通过媒体处理器 MediaPlayer 提供的方法，不仅可以播放存放在 SD 卡上音频文件，而且还能播放资源中的音频文件。这二者之间在设计方法上稍有不同。

下面按上述两种情况来说明应用 MediaPlayer 对象播放音频文件的步骤。

1. 创建 MediaPlayer 对象

（1）使用 new 方式创建 MediaPlayer 对象。

播放 SD 卡上的音频文件需要使用 new 方式创建 MediaPlayer 对象：

图 4.6 MediaPlayer 对象的生命周期

```
MediaPlayer mplayer = new MediaPlayer();
```

（2）使用 create()方法创建 MediaPlayer 对象。

播放资源中的音频需要使用 create 方法创建 MediaPlayer 对象，例如：

```
MediaPlayer mplayer = MediaPlayer.create(this, R.raw.test);
```

其中，R.raw.test 为资源中的音频数据源，test 为音频文件名称，注意不要带扩展名。

由于 create()方法中已经封装了初始化及同步的方法，故使用 create()方法创建的 MediaPlayer 对象不需要再进行初始化及同步操作。

2．设置播放文件

MediaPlayer 要播放的文件主要包括以下 3 个来源。

（1）存储在 SD 卡或其他文件路径下的音频文件

对于存储在 SD 卡中或其他文件路径下的音频文件，需要调用 setDataSource()方法，例如：

```
mplayer.setDataSource("/sdcard/test.mp3");
```

（2）在编写应用程序时事先存放在 res 资源中的音频文件。

对于事先存放在资源目录 res\raw 中的音频文件，在使用 create()方法创建 MediaPlayer 对象时，需要指定资源路径和文件名称（不要带扩展名）。由于 create()方法的源代码中已经封装了调用 setDataSource()方法，因此不必重复使用 setDataSource()方法。

(3)网络上的音频文件。

播放网络上的音频文件需要调用 setDataSource()方法,例如:

```
mplayer.setDataSource("http://www.citynorth.cn/music/confucius.mp3");
```

3. 对播放器进行同步控制

使用 prepare()方法设置对播放器的同步控制,例如:

```
mplayer.prepare();
```

如果 MediaPlayer 对象是由 create()方法创建的,由于 create()方法的源代码中已经封装了调用 prepare()方法,因此可省略此步骤。

4. 播放音频文件

start()是真正启动音频文件播放的方法,例如:

```
mplayer.start();
```

如要暂停播放或停止播放,则调用 pause()方法和 stop()方法。

5. 释放占用资源

音频文件播放结束时,应该调用 release()释放播放器占用的系统资源。

如要重新播放音频文件,需要调用 reset()返回到空闲状态,再从初始化开始重复其他各步骤。

【例 4-6】 设计一个简单的音乐播放器(音频文件存放在项目资源中)。

(1)创建应用程序项目,在资源目录 res 下新建 raw 目录,该目录存放比较大的音频、视频、图片等文件。将测试的音频文件 mtest.mp3 复制到 res\raw 目录下。

(2)设计界面布局。

在布局界面中,设置"开始播放"及"停止播放"按钮,如图 4.7 所示。

视频讲解

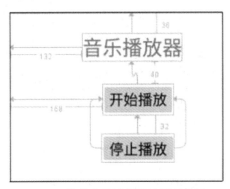

图 4.7 简单音乐播放器的界面布局

(3)设计控制程序 MainActivity.java。

```
1    package com.example.ex4_6;
```

```java
2   import androidx.appcompat.app.AppCompatActivity;
3   import android.os.Bundle;
4   import android.media.MediaPlayer;
5   import android.view.View;
6   import android.widget.Button;

7   public class MainActivity extends AppCompatActivity {
8     Button playBtn,  stopBtn;
9     MediaPlayer mMediaPlayer;
10    @Override
11    protected void onCreate(Bundle savedInstanceState) {
12      super.onCreate(savedInstanceState);
13      setContentView(R.layout.activity_main);
14      playBtn = (Button)findViewById(R.id.button);
15      stopBtn = (Button)findViewById(R.id.button2);
16      playBtn.setOnClickListener(new mPlay());   //设置监听对象
17      stopBtn.setOnClickListener(new mStop());
18    }
19    class mPlay implements View.OnClickListener{
20      @Override
21      public void onClick(View view) {
22        try{
23          mMediaPlayer = MediaPlayer.create(MainActivity.this, R.raw.mtest);
24          mMediaPlayer.start();    //开始播放
25        }catch (Exception e){   }
26      }
27    }

28    class  mStop  implements View.OnClickListener{
29      @Override
30      public void onClick(View view) {
31        try{
32          if(mMediaPlayer.isPlaying())
33          {
34            mMediaPlayer.stop();          //停止
35            mMediaPlayer.reset();         //重置
36            mMediaPlayer.release();       //释放所有占用资源
37          }
38        }catch (Exception e){   }
39      }
40    }

41  }
```

视频讲解

【**例 4-7**】 设计一个音乐播放器（音频文件存放在 SD 卡中）。

（1）将音频文件 **mtest.mp3** 复制到 SD 卡中（在模拟器中使用 SD 卡，可以在 Android Studio 集成环境中选择 Device File Explorer 调试工具，将音频文件复制到模拟器的 sdcard\

Music 目录下，如图 4.8 所示）。

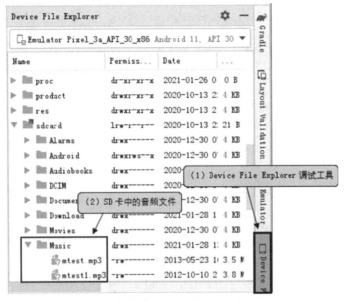

图 4.8　将音频文件存放到模拟器的 SD 卡内

（2）设计界面布局。

在用户界面布局中设置 3 个按钮，分别表示播放、暂停、停止，再设置一个进度条 SeekBar 组件，用于显示播放的进度，其界面布局如图 4.9 所示。

图 4.9　带进度条的音乐播放器的界面布局

（3）设计控制程序 MainActivity.java。

```
1   package com.example.ex4_7;
2   import androidx.appcompat.app.AppCompatActivity;
3   import android.os.Bundle;
4   import android.os.Build;
5   import android.Manifest;
6   import android.content.pm.PackageManager;
7   import android.media.MediaPlayer;
8   import android.util.Log;
```

```
 9    import android.view.View;
10    import android.widget.Button;
11    import android.widget.SeekBar;

12    public class MainActivity extends AppCompatActivity
13    {
14        Button btn_play, btn_pause, btn_stop;
15        MediaPlayer mplayer = new MediaPlayer();
16        String sdcard_file = "/sdcard/music/mtest.mp3";
17        SeekBar seekBar;

18        @Override
19        protected void onCreate(Bundle savedInstanceState)
20        {
21            super.onCreate(savedInstanceState);
22            setContentView(R.layout.activity_main);
23            btn_play=(Button)findViewById(R.id.button);
24            btn_pause=(Button)findViewById(R.id.button2);
25            btn_stop=(Button)findViewById(R.id.button3);
26            seekBar = (SeekBar)findViewById(R.id.seekBar);
27            getRegist();    //动态获取本地存储卡（SD 卡）读写权限操作
28            btn_play.setOnClickListener(new mStartClick());
29            btn_pause.setOnClickListener(new mPause());
30            btn_stop.setOnClickListener(new mStop());
31            seekBar.setOnSeekBarChangeListener(new mSeekBar()); //监听进度条操作
32        }

33        class mPause implements  View.OnClickListener
34        {
35            @Override
36            public void onClick(View v) {
37              try {
38                if (mplayer.isPlaying()) {
39                  mplayer.pause();
40                  btn_pause.setText("播放");
41                } else {
42                  mplayer.start();
43                  new myThread().start();
44                  btn_pause.setText("暂停");
45                }
46            }catch (Exception e){   }
47         }
48        }

49        class mStop implements View.OnClickListener
50        {
51          @Override
```

```
52    public void onClick(View v) {
53      try{
54        if(mplayer.isPlaying())
55        {
56          mplayer.stop();          //停止
57          mplayer.reset();         //重置
58          mplayer.release();       //释放资源
59        }
60      }catch(Exception e){ }
61    }
62  }

63  class mStartClick implements  View.OnClickListener
64  {
65    @Override
66    public void onClick(View v) {
67      try {
68        mplayer.setDataSource(sdcard_file);
69        playMusic(sdcard_file);
70      }catch (Exception e){ }
71    }
72  }

73  private void playMusic(String path)
74  {
75    try {
76      mplayer.reset();;
77      mplayer.setDataSource(path);
78      mplayer.prepare();
79      seekBar.setMax(mplayer.getDuration());
80      mplayer.start();
81      new myThread().start();                  //用多线程控制进度条显示
82    }catch (Exception e){ }
83  }

84  //动态获取本地存储卡（SD卡）读写权限
85  private void getRegist()
86  {
87    /*
88        动态获取一些保护权限，如文件读写
89        除了要在AndroidManifest中声明权限之外，还要使用如下代码动态获取
90    */
91    if (Build.VERSION.SDK_INT >= 23){  //23为API版本Android 6.0
92      int REQUEST_CODE_CONTACT = 101;
93      final int REQUEST_EXTERNAL_STORAGE = 1;
94      String[] PERMISSIONS_STORAGE = {
95        Manifest.permission.READ_EXTERNAL_STORAGE,
```

```
96              Manifest.permission.WRITE_EXTERNAL_STORAGE
97          };
98      //验证是否许可权限
99      for (String str : PERMISSIONS_STORAGE) {
100         if (this.checkSelfPermission(str) !=
101             PackageManager.PERMISSION_GRANTED){
102         //申请权限
103         this.requestPermissions(PERMISSIONS_STORAGE, REQUEST_CODE_CONTACT);
104         return;
105         }
106     }
107 }
108 }

109 // 进度条操作
110 class mSeekBar implements  SeekBar.OnSeekBarChangeListener
111 {
112     @Override
113     public void onProgressChanged(SeekBar seekBar,
114                 int progress, boolean fromUser) {
115         mplayer.seekTo(progress);
116     }
117     @Override
118     public void onStartTrackingTouch(SeekBar seekBar) { }
119     @Override
120     public void onStopTrackingTouch(SeekBar seekBar) {    }
121 }

122     class myThread extends Thread
123     {
124       public void run() {
125         while (true) {
126           seekBar.setProgress(mplayer.getCurrentPosition());
127           try {
128             Thread.sleep(1000);
129           } catch (Exception e) {
130             e.printStackTrace();
131           }
132         }
133       }
134 }

135 }
```

(4) 修改配置文件，添加存储卡的读写权限。

```
1   <?xml version="1.0" encoding="utf-8"?>
```

```
2    <manifest xmlns:android="http://schemas.android.com/apk/res/android"
3        package="com.example.ex4_7">
4        <application
5            android:allowBackup="true"
6            android:icon="@mipmap/ic_launcher"
7            android:label="@string/app_name"
8            android:roundIcon="@mipmap/ic_launcher_round"
9            android:supportsRtl="true"
10           android:theme="@style/Theme.ex4_7">
11           <activity android:name=".MainActivity">
12               <intent-filter>
13                   <action android:name="android.intent.action.MAIN" />
14                   <category android:name="android.intent.category.LAUNCHER" />
15               </intent-filter>
16           </activity>
17       </application>
18       <uses-permission android:name="android.permission.WRITE_EXTERNAL_STORAGE" />
19       <uses-permission android:name="android.permission.READ_EXTERNAL_STORAGE" />
20   </manifest>
```

4.4 视频播放

在 Android 系统中，设计播放视频的应用程序有两种方式，一种方式是应用媒体播放器组件 MediaPlayer 播放视频，另一种方式是应用视频视图组件 VideoView 播放视频。下面分别介绍这两种设计方式。

4.4.1 应用媒体播放器播放视频

应用媒体播放器 MediaPlayer 不仅可以播放音频文件，而且还可以播放视频文件。与播放音频有所不同，用于视频播放的播放承载体必须是实现了表面视图处理接口（surfaceHolder）的视图组件，即需要使用 SurfaceView 组件来显示播放的视频图像。

【例4-8】 应用媒体播放器组件 MediaPlayer 设计一个视频播放器。

（1）事先准备视频文件 sample.mp4，并将其复制到新建的 res\raw 目录下。

（2）设计布局文件 activity_main.xml。

在用户界面布局中，设置一个 SurfaceView 组件，用于显示视频图像；再设置两个按钮，用于播放视频和停止播放。布局文件的源代码如下：

视频讲解

```
1    <?xml version="1.0" encoding="utf-8"?>
```

```
2    <androidx.constraintlayout.widget.ConstraintLayout
3    xmlns:android="http://schemas.android.com/apk/res/android"
4      xmlns:app="http://schemas.android.com/apk/res-auto"
5      xmlns:tools="http://schemas.android.com/tools"
6      android:layout_width="match_parent"
7      android:layout_height="match_parent"
8      tools:context=".MainActivity">

9    <LinearLayout
10     android:id="@+id/linearLayout"
11     android:layout_width="366dp"
12     android:layout_height="474dp"
13     android:layout_marginStart="8dp"
14     android:layout_marginLeft="8dp"
15     android:layout_marginTop="8dp"
16     android:orientation="vertical"
17     app:layout_constraintStart_toStartOf="parent"
18     app:layout_constraintTop_toTopOf="parent">

19     <SurfaceView              ◄──── 用于显示视频图像
20       android:id="@+id/surfaceView"
21       android:layout_width="354dp"
22       android:layout_height="271dp" />

23     <LinearLayout
24       android:layout_width="match_ parent"
25       android:layout_height="49dp"
26       android:gravity="center"
27       android:orientation="horizontal">

28       <Button
29         android:id="@+id/button"
30         android:layout_width="122dp"
31         android:layout_height="47dp"
32         android:text="播放" />

33       <Button
34         android:id="@+id/button2"
35         android:layout_width="wrap_content"
36         android:layout_height="wrap_content"
37         android:text="停止" />
38     </LinearLayout>
39   </LinearLayout>

40   </androidx.constraintlayout.widget.ConstraintLayout>
```

（3）设计控制程序 MainActivity.java。

```java
1   package com.example.ex4_8;
2   import androidx.appcompat.app.AppCompatActivity;
3   import android.os.Bundle;

4   import android.util.Log;
5   import android.view.SurfaceHolder;
6   import android.view.SurfaceView;
7   import android.view.View;
8   import android.widget.Button;
9   import android.content.res.AssetFileDescriptor;
10  import android.media.AudioManager;
11  import android.media.MediaPlayer;

12  public class MainActivity extends AppCompatActivity
13  {
14      MediaPlayer mMediaPlayer;
15      SurfaceView mSurfaceView;
16      private Button playBtn, stopBtn;
17      SurfaceHolder sh;
18      @Override
19      protected void onCreate(Bundle savedInstanceState)
20      {
21          super.onCreate(savedInstanceState);
22          setContentView(R.layout.activity_main);
23          mSurfaceView = (SurfaceView) findViewById(R.id.surfaceView);
24          playBtn = (Button) findViewById(R.id.button);
25          stopBtn = (Button)findViewById(R.id.button2);
26          playBtn.setOnClickListener(new mplay());
27          stopBtn.setOnClickListener(new mstop());
28      }

29      class mplay implements View.OnClickListener
30      {
31          @Override
32          public void onClick(View v) {
33          try {
34            mMediaPlayer = new MediaPlayer();
35            mMediaPlayer.reset();
36            //为播放器对象设置用于显示视频内容、代表屏幕描绘的控制器
37            mMediaPlayer.setAudioStreamType(AudioManager.STREAM_MUSIC);
38            AssetFileDescriptor assetFileDescriptor =
39              getResources().openRawResourceFd(R.raw.sample);   //设置数据源
40            mMediaPlayer.setDataSource(
41                assetFileDescriptor.getFileDescriptor(),
42                assetFileDescriptor.getStartOffset(),
43                assetFileDescriptor.getLength()
```

```
44            );
45            sh=mSurfaceView.getHolder();
46            mMediaPlayer.setDisplay(sh);
47            mMediaPlayer.prepare();
48            mMediaPlayer.start();
49        } catch (Exception e) {
50            Log.i("MediaPlay err", "MediaPlay err");
51        }
52    }
53  }

54  class mstop implements View.OnClickListener
55  {
56    @Override
57    public void onClick(View v) {
58        mMediaPlayer.stop();
59        mMediaPlayer.release();
60    }
61  }
62 }
```

视频播放程序的运行结果如图 4.10 所示。

图 4.10　应用媒体播放器播放视频

■ 4.4.2　应用视频视图播放视频

在 Android 系统中，经常使用 android.widget 包中的视频视图类 VideoView 播放视频文件。VideoView 类可以从不同的来源（如资源文件或内容提供器）读取图像，计算和维护视频的画面尺寸以使其适用于任何布局管理器，并提供一些诸如缩放、着色之类的显示选项。VideoView 类的常用方法如表 4-6 所示。

第4章 图形与多媒体处理

表 4-6 VideoView 类的常用方法

方法	说明
VideoView(Context context)	创建一个默认属性的 VideoView 实例
boolean canPause()	判断是否能够暂停播放视频
int getBufferPercentage()	获得缓冲区的百分比
int getCurrentPosition()	获得当前的位置
int getDuration()	获得所播放视频的总时间
boolean isPlaying()	判断是否正在播放视频
boolean onTouchEvent(MotionEvent ev)	实现该方法来处理触屏事件
seekTo(int msec)	设置播放位置
setMediaController(MediaController controller)	设置媒体控制器
setOnCompletionListener(MediaPlayer.OnCompletionListener l)	注册在媒体文件播放完毕时调用的回调函数
setOnPreparedListener(MediaPlayer.OnPreparedListener l)	注册在媒体文件加载完毕并可以播放时调用的回调函数
setVideoPath(String path)	设置视频文件的路径名
setVideoURI(Uri uri)	设置视频文件的统一资源标识符
start()	开始播放视频文件
stopPlayback()	停止回放视频文件

【例 4-9】 应用视频视图组件 VideoView 设计一个视频播放器。

(1) 事先准备视频文件 led.mp4，并将其复制到新建的 res\raw 目录下。

(2) 设计布局文件 activity_main.xml。

在用户界面布局中，设置一个 VideoView 组件，用于显示视频图像；再设置两个按钮，用于播放视频和暂停播放。布局文件的源代码如下：

```xml
1   <?xml version="1.0" encoding="utf-8"?>
2   <androidx.constraintlayout.widget.ConstraintLayout
3       xmlns:android="http://schemas.android.com/apk/res/android"
4       xmlns:app="http://schemas.android.com/apk/res-auto"
5       xmlns:tools="http://schemas.android.com/tools"
6       android:layout_width="match_parent"
7       android:layout_height="match_parent"
8       tools:context=".MainActivity">

9       <LinearLayout
10          android:id="@+id/linearLayout"
11          android:layout_width="366dp"
12          android:layout_height="474dp"
13          android:layout_marginStart="8dp"
14          android:layout_marginLeft="8dp"
15          android:layout_marginTop="8dp"
16          android:orientation="vertical"
17          app:layout_constraintStart_toStartOf="parent"
18          app:layout_constraintTop_toTopOf="parent">
```

```
19    <VideoView                    ←── 用于显示视频图像
20      android:id="@+id/videoView"
21      android:layout_width="354dp"
22      android:layout_height="271dp" />

23    <LinearLayout
24      android:layout_width="match_parent"
25      android:layout_height="49dp"
26      android:gravity="center"
27      android:orientation="horizontal">

28      <Button
29        android:id="@+id/button"
30        android:layout_width="122dp"
31        android:layout_height="47dp"
32        android:text="播放" />

33      <Button
34        android:id="@+id/button2"
35        android:layout_width="wrap_content"
36        android:layout_height="wrap_content"
37        android:text="暂停" />
38    </LinearLayout>
39  </LinearLayout>

40  </androidx.constraintlayout.widget.ConstraintLayout>
```

（3）设计控制文件 MainActivity.java。

```
1   package com.example.ex4_9;
2   import androidx.appcompat.app.AppCompatActivity;
3   import android.os.Bundle;
4   import android.net.Uri;
5   import android.view.View;
6   import android.widget.Button;
7   import android.widget.MediaController;
8   import android.widget.VideoView;

9   public class MainActivity extends AppCompatActivity
10  {
11      private VideoView mVideoView;
12      private Button playBtn, pauseBtn;
13      MediaController mMediaController;
14      @Override
15      protected void onCreate(Bundle savedInstanceState)
16      {
17          super.onCreate(savedInstanceState);
18          setContentView(R.layout.activity_main);
19          mVideoView = (VideoView)findViewById(R.id.videoView);
20          playBtn = (Button)findViewById(R.id.button);
```

```
21        pauseBtn = (Button)findViewById(R.id.button2);
22        init();
23        playBtn.setOnClickListener(new mPlay());
24        pauseBtn.setOnClickListener(new mPause());
25    }

26    void init(){
27        String uri = "android.resource://" + getPackageName()
28                    + "/" + R.raw.led;                          //设置视频源
29        mVideoView.setVideoURI(Uri.parse(uri));
30        mMediaController = new MediaController(this);//实例化媒体控制器
31        mMediaController.setMediaPlayer(mVideoView);
32        mVideoView.setMediaController(mMediaController);
33    }

34    class mPlay implements View.OnClickListener{
35        @Override
36        public void onClick(View v) {
37            mVideoView.start();
38        }
39    }

40    class mPause implements View.OnClickListener{
41        @Override
42        public void onClick(View v) {
43            if(mVideoView.isPlaying()) {
44                mVideoView.pause();
45            }
46        }
47    }

48 }
```

程序运行结果如图 4.11 所示。

图 4.11 应用视频视图（VideoView）播放视频

4.5 文本转换语音

Android 系统自带的 TextToSpeech 组件可以将一段文字转换为语音，还可根据需要合成出不同音色、语速和语调的声音，让机器像人一样开口说话。

1. TextToSpeech 组件的常用方法

TextToSpeech 组件的常用方法如表 4-7 所示。

表 4-7 TextToSpeech 组件的常用方法

方法	说明
setLanguage(Locale)	设置语言种类，设置中文为 Locale.CHINA
setPitch(float)	设置音调，值越大声音越尖（女声），值越小声音越粗（男声），1.0 是默认值
setSpeechRate(float)	设定语速，默认值 1.0 为正常语速
speak()	朗读文本（将文字转换为语音，并朗读出来）
stop()	停止朗读
shutdown()	释放占用的资源
isSpeaking()	是否正在转换

2. TextToSpeech 的初始化监听接口

TextToSpeech 组件有一个用于初始化的监听接口 OnInitListener，需要重新定义其 onInit(int status) 方法，在该方法中进行初始化设置。

【例 4-10】 应用 TextToSpeech 组件，将文字转换为语音。

（1）设计界面布局。

在界面布局中，设置一个文本编辑框 EditText，用于输入文字内容；再设置一个按钮，将文字内容转换成语音输出。

（2）设计控制程序。

```
1    package com.example.ex4_10;
2    import androidx.appcompat.app.AppCompatActivity;
3    import android.os.Bundle;
4    import android.speech.tts.TextToSpeech;
5    import android.view.View;
6    import android.widget.Button;
7    import android.widget.EditText;
8    import android.widget.Toast;
9    import java.util.Locale;
10   public class MainActivity extends AppCompatActivity
11   {
12       private Button speechBtn;           // 按钮控制开始朗读
```

```
13    private EditText speechTxt;     // 需要朗读的内容
14    private TextToSpeech textToSpeech; // TTS 对象
15    @Override
16    protected void onCreate(Bundle savedInstanceState) {
17      super.onCreate(savedInstanceState);
18      setContentView(R.layout.activity_main);
19      speechBtn = (Button) findViewById(R.id.button);
20      speechTxt = (EditText) findViewById(R.id.editText);
21      textToSpeech = new TextToSpeech(MainActivity.this, new mClick());
22      speechBtn.setOnClickListener(new mClick());
23    }

24    class mClick implements View.OnClickListener, TextToSpeech.OnInitListener
25    {
26      /**
27       初始化 TextToSpeech 引擎
28       status:取值为 SUCCESS 或 ERROR
29       setLanguage:设置语言,帮助文档里有 22 种
30       TextToSpeech.LANG_MISSING_DATA:表示语言的数据丢失
31       TextToSpeech.LANG_NOT_SUPPORTED:表示语言的数据不支持
32      */
33      @Override
34      public void onInit(int status) {
35        if (status == TextToSpeech.SUCCESS)
36        {
37          int result = textToSpeech.setLanguage(Locale.CHINA);
38          if (result == TextToSpeech.LANG_MISSING_DATA
39              || result == TextToSpeech.LANG_NOT_SUPPORTED)
40          {
41            Toast.makeText(this, "数据丢失或不支持",
42                Toast.LENGTH_SHORT).show();
43          }
44        }
45      }

46      @Override
47      public void onClick(View v)
48      {
49        if (textToSpeech != null && !textToSpeech.isSpeaking())
50        {
51          textToSpeech.setPitch(0.5f);            //设置音调
52          textToSpeech.setSpeechRate(1.2f);       //设定语速
53          textToSpeech.speak(speechTxt.getText().toString(),
54          TextToSpeech.QUEUE_FLUSH,  null);       //朗读
55        }
56     }
```

```
57      @Override
58      protected void onStop()
59      {
60        super.onStop();
61        textToSpeech.stop();          // 不管是否正在朗读 TTS 都被中断
62        textToSpeech.shutdown();      // 关闭，释放资源
63      }
64  }
```

程序运行结果如图 4.12 所示，在编辑框中输入文本字符，点击"转换语音"按钮，将语音朗读出来。

图 4.12　将文字内容转换为语音

 4.6　动画技术

■ 4.6.1　动画组件类

1. 动画组件概述

动画组件（Animation）是一个实现 Android UI 界面动画效果的 API，Animation 提供了一系列的动画效果，可以进行旋转、缩放、淡入淡出等，这些效果可以应用在绝大多数的控件中。

2. 动画组件的分类

Animation 从总体上可以分为以下三大类：

（1）补间动画。

补间动画（Tween Animation）只需指定开始、结束的"关键帧"，而变化中的其他帧由系统来计算，不必一帧一帧地去定义。

（2）逐帧动画。

逐帧动画（Frame Animation）可以创建一个 Drawable 序列，这些 Drawable 可以按照

指定的时间间歇逐一显示。

（3）属性动画。

属性动画（Property Animation）是在 Android 3.0 版本以后才引进的，它可以直接更改对象的属性。在上面提到的 Tween Animation 中，只是更改 View 的绘画效果，而 View 的真实属性是不改变的。假设用 Tween Animation 动画将一个 Button 从左边移到右边，无论怎么点击移动后的 Button 都没有反应，而当点击移动前 Button 的位置时才有反应，因为 Button 的位置属性没有改变。Property Animation 属性动画则可以直接改变 View 对象的属性值，这样可以让编程人员少做一些处理工作，提高效率与代码的可读性。

■ 4.6.2 补间动画

1．补间动画效果的种类及对应的子类

补间动画共有 4 种动画效果及对应的子类。
（1）Alpha：淡入淡出效果，其对应子类为 AlphaAnimation。
（2）Scale：缩放效果，其对应子类为 ScaleAnimation。
（3）Rotate：旋转效果，其对应子类为 RotateAnimation。
（4）Translate：移动效果，其对应子类为 TranslateAnimation。
补间动画效果对应子类的构造方法如表 4-8 所示。

表 4-8　补间动画效果对应子类的构造方法

对应子类的构造方法	参　数　说　明
AlphaAnimation (　　float fromAlpha, 　　float toAlpha)	参数 1 fromAlpha：起始透明度 参数 2 toAlpha：终止透明度 （取 0.0~1.0 的数值，1.0 为完全不透明，0.0 为完全透明）
ScaleAnimation(　　float fromX, 　　float toX, 　　float fromY, 　　float toY, 　　int pivotXType, 　　float pivotXValue, 　　int pivotYType, 　　float pivotYValue)	参数 1 fromX：X 轴的初始值 参数 2 toX：X 轴收缩后的值 参数 3 fromY：Y 轴的初始值 参数 4 toY：Y 轴收缩后的值 参数 5 pivotXType：确定 X 轴坐标的类型 参数 6 pivotXValue：X 轴的值，0.5f 表明是以自身这个控件的一半长度为 X 轴 参数 7 pivotYType：确定 Y 轴坐标的类型 参数 8 pivotYValue：Y 轴的值，0.5f 表明是以自身这个控件的一半长度为 Y 轴
RotateAnimation(　　float fromDegrees, 　　float toDegrees, 　　int pivotXType, 　　float pivotXValue, 　　int pivotYType, 　　float pivotYValue)	参数 1 fromDegrees：从哪个旋转角度开始 参数 2 toDegrees：转到什么角度 　　后面 4 个参数用于设置围绕着旋转的圆的圆心位置 参数 3 pivotXType：确定 X 轴坐标的类型，有 ABSOLUT 绝对坐标、RELATIVE_TO_SELF 相对于自身坐标、RELATIVE_TO_PARENT 相对于父控件的坐标 参数 4 pivotXValue：X 轴的值，0.5f 表明是以自身这个控件的一半长度为 X 轴 参数 5 pivotYType：确定 Y 轴坐标的类型 参数 6 pivotYValue：Y 轴的值，0.5f 表明是以自身这个控件的一半长度为 Y 轴

续表

对应子类的构造方法	参数说明
TranslateAnimation(　　float fromXDelta, 　　float toXDelta, 　　float fromYDelta, 　　float toYDelta)	参数 1 fromXDelta：X 轴的开始位置 参数 2 toXDelta：X 轴的结束位置 参数 3 fromYDelta：Y 轴的开始位置 参数 4 toYDelta：Y 轴的结束位置

2. AnimationSet 类

AnimationSet 类是 Animation 的子类，用于设置 Animation 的属性。

【例 4-11】 编写一个可以旋转、缩放、淡入淡出、移动的补间动画程序。

（1）新建工程 ex4_11，并将事先准备的图片 an.jpg 复制到项目的 res\drawable 目录下，如图 4.13 所示。

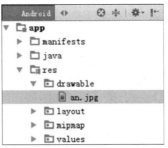

图 4.13　将图片 an.jpg 复制到项目的 res\drawable 目录下

（2）创建自定义组件。

创建一个显示动画的组件 AniView.java，其代码如下。

```
1    package com.example.ex4_11;
2    import android.content.Context;
3    import android.graphics.Bitmap;
4    import android.graphics.BitmapFactory;
5    import android.graphics.Canvas;
6    import android.graphics.Color;
7    import android.util.AttributeSet;
8    import android.view.View;

9    class AniView extends View
10   {
11       Bitmap img1,img2;
12       public AniView(Context context, AttributeSet attrs)
13       {
14           super(context, attrs);
15           img1 = BitmapFactory.decodeResource(getResources(), R.drawable.an);
16           img2 = Bitmap.createBitmap(img1);
17       }
```

```
18    protected void onDraw(Canvas canvas)
19    {
20      super.onDraw(canvas);
21      canvas.drawBitmap(img2, 0, 0, null);
22    }
23 }
```

（3）设计界面布局。

在界面布局程序中，设置 4 个按钮组件 Button 和一个自定义显示动画组件 AniView。界面布局程序的代码如下：

```
1  <?xml version="1.0" encoding="utf-8"?>
2  <androidx.constraintlayout.widget.ConstraintLayout
3   xmlns:android="http://schemas.android.com/apk/res/android"
4   xmlns:app="http://schemas.android.com/apk/res-auto"
5   xmlns:tools="http://schemas.android.com/tools"
6   android:layout_width="match_parent"
7   android:layout_height="match_parent"
8   tools:context=".MainActivity">

9   <TextView
10    android:id="@+id/textView"
11    android:layout_width="wrap_content"
12    android:layout_height="wrap_content"
13    android:text="动画演示"
14    android:textColor="@color/teal_700"
15    android:textSize="30sp"
16    android:textStyle="bold"
17    app:layout_constraintBottom_toBottomOf="parent"
18    app:layout_constraintHorizontal_bias="0.446"
19    app:layout_constraintLeft_toLeftOf="parent"
20    app:layout_constraintRight_toRightOf="parent"
21    app:layout_constraintTop_toTopOf="parent"
22    app:layout_constraintVertical_bias="0.053" />

23   <Button
24    android:id="@+id/rotateButton"
25    android:layout_width="69dp"
26    android:layout_height="52dp"
27    android:layout_marginStart="4dp"
28    android:layout_marginLeft="4dp"
29    android:layout_marginTop="16dp"
30    android:text="旋转"
31    app:layout_constraintStart_toStartOf="parent"
32    app:layout_constraintTop_toBottomOf="@+id/textView" />

33   <Button
```

```
34      android:id="@+id/scaleButton"
35      android:layout_width="68dp"
36      android:layout_height="48dp"
37      android:layout_marginStart="8dp"
38      android:layout_marginLeft="8dp"
39      android:text="缩放"
40      app:layout_constraintStart_toEndOf="@+id/rotateButton"
41      app:layout_constraintTop_toTopOf="@+id/rotateButton" />

42    <Button
43      android:id="@+id/alphaButton"
44      android:layout_width="93dp"
45      android:layout_height="46dp"
46      android:layout_marginStart="8dp"
47      android:layout_marginLeft="8dp"
48      android:text="淡入淡出"
49      app:layout_constraintStart_toEndOf="@+id/scaleButton"
50      app:layout_constraintTop_toTopOf="@+id/scaleButton" />

51    <Button
52      android:id="@+id/translateButton"
53      android:layout_width="72dp"
54      android:layout_height="48dp"
55      android:layout_marginStart="8dp"
56      android:layout_marginLeft="8dp"
57      android:text="移动"
58      app:layout_constraintStart_toEndOf="@+id/alphaButton"
59      app:layout_constraintTop_toTopOf="@+id/alphaButton" />

60    <com.example.ex4_11.AniView
61      android:id="@+id/rotateView"
62      android:layout_width="201dp"
63      android:layout_height="192dp"
64      android:layout_marginTop="16dp"
65      app:layout_constraintEnd_toEndOf="@+id/textView"
66      app:layout_constraintStart_toStartOf="@+id/textView"
67      app:layout_constraintTop_toBottomOf="@+id/translateButton" />

68   </androidx.constraintlayout.widget.ConstraintLayout>
```

这时，单击菜单 Build 中的 Rebuild Project 选项，重新编译应用程序，可以在编辑器的预览中看到所自定义的组件。

（4）编写控制层 MainActivity.java 程序。

```
1    package com.example.ex4_11;
2    import androidx.appcompat.app.AppCompatActivity;
3    import android.os.Bundle;
```

```java
4   import android.view.View;
5   import android.view.animation.AlphaAnimation;
6   import android.view.animation.Animation;
7   import android.view.animation.ScaleAnimation;
8   import android.view.animation.TranslateAnimation;
9   import android.view.animation.RotateAnimation;
10  import android.widget.Button;

11  public class MainActivity extends AppCompatActivity
12  {
13    private Button rotateButton = null;
14    private Button scaleButton = null;
15    private Button alphaButton = null;
16    private Button translateButton = null;
17    RotateAnimation rotateAnimation;
18    AniView rotateView;

19    @Override
20    protected void onCreate(Bundle savedInstanceState)
21    {
22      super.onCreate(savedInstanceState);
23      setContentView(R.layout.activity_main);
24      rotateButton = (Button)findViewById(R.id.rotateButton);
25      scaleButton = (Button)findViewById(R.id.scaleButton);
26      alphaButton = (Button)findViewById(R.id.alphaButton);
27      translateButton = (Button)findViewById(R.id.translateButton);
28      rotateView = (AniView)findViewById(R.id.rotateView);
29      rotateButton.setOnClickListener(new RotateButtonListener());
30      scaleButton.setOnClickListener(new ScaleButtonListener());
31      alphaButton.setOnClickListener(new AlphaButtonListener());
32      translateButton.setOnClickListener(new TranslateButtonListener());
33    }
34    class RotateButtonListener implements OnClickListener{
35      public void onClick(View v) {
36        RotateAnimation rotateAnimation = new RotateAnimation(0, 360,
37             Animation.RELATIVE_TO_SELF, 0.5f,
38             Animation.RELATIVE_TO_SELF, 0.5f);
39        rotateAnimation.setDuration(1000);
40        rotateView.startAnimation(rotateAnimation);
41      }
42    }
43    class ScaleButtonListener implements OnClickListener{
44      public void onClick(View v) {
45        ScaleAnimation scaleAnimation = new ScaleAnimation(
46             0, 0.1f, 0, 0.1f,
47             Animation.RELATIVE_TO_SELF, 0.5f,
48             Animation.RELATIVE_TO_SELF, 0.5f );
```

```
49          scaleAnimation.setDuration(1000);
50          rotateView.startAnimation(scaleAnimation);
51      }
52  }
53  class AlphaButtonListener implements OnClickListener{
54      public void onClick(View v) {
55          AlphaAnimation alphaAnimation = new AlphaAnimation(1, 0);
56          alphaAnimation.setDuration(500);
57          rotateView.startAnimation(alphaAnimation);
58      }
59  }
60  class TranslateButtonListener implements OnClickListener{
61      public void onClick(View v) {
62          TranslateAnimation translateAnimation =
63              new TranslateAnimation(
64                  Animation.RELATIVE_TO_SELF, 0f,
65                  Animation.RELATIVE_TO_SELF, 0.5f,
66                  Animation.RELATIVE_TO_SELF, 0f,
67                  Animation.RELATIVE_TO_SELF, 0.5f);
68          translateAnimation.setDuration(1000);
69          rotateView.startAnimation(translateAnimation);
70      }
71  }
72  }
```

程序运行结果如图 4.14 所示。

图 4.14 补间动画示例

4.6.3 属性动画

1. 属性动画的核心类

属性动画是通过控制对象中的属性值产生的动画,主要的核心类有 ValueAnimator 和 ObjectAnimator。

(1) ValueAnimator 类。

ValueAnimator 是整个属性动画机制当中最核心的一个类。属性动画的运行机制是通过不断地对值进行操作来实现的,而初始值和结束值之间的动画过渡是由 ValueAnimator 这个类来负责计算的。它的内部使用一种时间循环的机制来计算值与值之间的动画过渡,用户只需要将初始值和结束值提供给 ValueAnimator,并且告诉它动画所需运行的时长,那么 ValueAnimator 就会自动帮助用户完成从初始值平滑地过渡到结束值这样的效果。除此之外,ValueAnimator 还负责管理动画的播放次数、播放模式以及对动画设置监听器等。

(2) ObjectAnimator 类。

ObjectAnimator 是 ValueAnimator 的子类,它本身就已经包含了时间引擎和值计算,所以它拥有为对象的某个属性设置动画的功能,这使得为任何对象设置动画更加的容易。

ObjectAnimator 类是设计动画中最常使用的类,ValueAnimator 只不过是对值进行了一个平滑的动画过渡,所以实际使用到这种功能的场景并不多。而 ObjectAnimator 则不同,它可以直接对任意对象的任意属性进行动画操作,比如 View 的 alpha 属性。

构造 ObjectAnimator 对象的方法为 ofFloat(),其方法原型如下:

```
public static ObjectAnimator ofFloat(
    Object target,
    String propertyName,
    float... values
    );
```

- 第 1 个参数用于指定动画对象要操作的控件。
- 第 2 个参数用于指定动画对象所要操作控件的属性。
- 第 3 个参数是可变长参数,设置动画的起点和终点位置。

2. ObjectAnimator 类实现动画示例

下面通过一个示例来说明如何应用 ObjectAnimator 类实现动画。

【例 4-12】 编写一个可以旋转、缩放、淡入淡出的属性动画程序。

(1) 新建应用程序 ex4_12,并将事先准备的图片 an.jpg 复制到 res\drawable 目录下。
(2) 设计界面布局。

在界面布局程序中,放置 3 个按钮组件 Button,分别设置为"旋转""缩放""淡入淡出";再设置一个 ImageView 组件,用于显示动画。

(3) 编写控制层 Java 程序。

```
1    package com.example.ex4_12;
```

```java
2    import androidx.appcompat.app.AppCompatActivity;
3    import android.os.Bundle;
4    import android.widget.Button;
5    import android.widget.ImageView;
6    import android.view.View;
7    import android.animation.ObjectAnimator;

8    public class MainActivity extends AppCompatActivity
9    {
10       Button rotateButton,alphaButton,scaleButton;
11       ObjectAnimator animator;
12       ImageView img;

13       @Override
14       protected void onCreate(Bundle savedInstanceState)
15       {
16         super.onCreate(savedInstanceState);
17         setContentView(R.layout.activity_main);
18         img = (ImageView)findViewById(R.id.imageView);
19         rotateButton = (Button)findViewById(R.id.button1);
20         alphaButton = (Button)findViewById(R.id.button2);
21         scaleButton = (Button)findViewById(R.id.button3);
22         rotateButton.setOnClickListener(new mClick());
23         alphaButton.setOnClickListener(new mClick());
24         scaleButton.setOnClickListener(new mClick());
25       }
26       public class mClick implements View.OnClickListener
27       {
28          @Override
29          public void onClick(View v)
30          {
31            if(v == rotateButton)
32             {
33               animator = ObjectAnimator.ofFloat(img,
34                       "rotation", 0.0F, 360.0F);
35               animator.setDuration(1000);
36               animator.start();
37             }
38            else if(v == alphaButton){
39               animator = ObjectAnimator.ofFloat(img,
40                       "alpha",1.0F, 0.0F, 1.0F);
41               animator.setDuration(3000);
42               animator.start();
43             }
44            else if(v == scaleButton){
45               animator = ObjectAnimator.ofFloat(img,
46              "ScaleY", 1.0F, 0.5F, 1.0F);
```

```
47              animator.setDuration(5000);
48              animator.start();
49          }
50      }
51  }
52  }
```

程序运行结果如图 4.15 所示。

图 4.15　属性动画示例

习题 4

1. 设计一个可以移动的小球，当小球被拖到一个小矩形块中时退出程序。
2. 设计一个手绘图形的画板。
3. 设计一个具有选歌功能的音频播放器。
4. 为例 4-9 的视频播放器添加停止播放的功能。
5. 编写一个具有飞入文字功能的程序。

第5章　后台服务与系统服务

5.1 后台服务

Android 系统的后台服务（Service）是一种类似于 Activity 的组件，但 Service 没有用户操作界面，也不能自己启动，其主要作用是提供后台服务调用。Service 不像 Activity 那样当用户关闭应用界面就停止运行，Service 会一直在后台运行，除非明确命令其停止。

通常使用 Service 为应用程序提供一些只需在后台运行的服务或不需要界面的功能，如从 Internet 下载文件、控制 Video 播放器等。

Service 的生命周期中只有 3 个阶段，即 onCreate、onStartCommand、onDestroy。Service 的常用方法如表 5-1 所示。

表 5-1　Service 的常用方法

方　　法	说　　明
onCreate()	创建后台服务
onStartCommand (Intent intent, int flags, int startId)	启动后台服务
onDestroy()	销毁后台服务，并删除所有调用
sendBroadcast(Intent intent)	继承父类 Context 的 sendBroadcast()方法，实现发送广播机制的消息
onBind(Intent intent)	与服务通信的信道进行绑定，服务程序必须实现该方法
onUnbind(Intent intent)	撤销与服务信道的绑定

通常 Service 要在一个 Activity 中启动，调用 Activity 的 startService(Intent)方法启动 Service。若要停止正在运行的 Service，则调用 Activity 的 stopService(Intent)方法关闭 Service。startService() 和 stopService()方法均继承于 Activity 及 Service 共同的父类 android.content.Context。

一个服务只能创建一次，销毁一次，但可以开始多次，即 onCreate()和 onDestroy()方法只会被调用一次，而 onStartCommand()方法可以被调用多次。后台服务的具体操作一般应该放在 onStartCommand()方法里面。如果 Service 已经启动，当再次启动 Service 时则不调用 onCreate()而直接调用 onStartCommand()。

设计一个后台服务的应用程序大致有以下几个步骤。

（1）创建 Service 的子类。

① 编写 onCreate()方法，创建后台服务；

② 编写 onStartCommand()方法，启动后台服务；

③ 编写 onDestroy()方法，终止后台服务，并删除所有调用。

（2）创建启动和控制 Service 的 Activity。

① 创建 Intent 对象，建立 Activity 与 Service 的关联；

② 调用 Activity 的 startService(Intent)方法启动 Service 后台服务；

③ 调用 Activity 的 stopService(Intent)方法关闭 Service 后台服务。

（3）修改配置文件 AndroidManifest.xml。

在配置文件 AndroidManifest.xml 的<application>标签中添加以下代码：

```
<service android:enabled="true" android:name=".AudioSrv" />
```

【例 5-1】 一个简单的后台音频服务程序示例。

本例通过一个按钮启动后台服务，在服务程序中播放音频文件，演示服务程序的创建、启动，再通过另一个按钮演示服务程序的销毁过程。新建项目 ex5_1 后，将音频文件 mtest.mp3 复制到应用程序新建的 res\raw 目录下。

（1）设计界面布局。

在界面布局中，设置两个按钮，分别用于"启动后台服务程序"和"关闭后台服务程序"；再设置一个文本标签，用于显示后台服务程序的运行状态。

（2）设计后台服务程序。

新建一个后台服务程序，命名为 AudioSrv.java，其代码如下。

```
1   package com.example.ex5_1;
2   import android.app.Service;
3   import android.content.Intent;
4   import android.media.MediaPlayer;
5   import android.os.IBinder;
6   import android.widget.Toast;
7
8   public class AudioSrv extends Service
9   {
10      MediaPlayer play;
11      @Override
12      public IBinder onBind(Intent intent)
13      {
14          return null;
15      }
16      public void onCreate()
17      {
18          super.onCreate();
19          play = MediaPlayer.create(this, R.raw.mtest);   ◀创建调用资源音频文件对象
20          Toast.makeText(this, "创建后台服务...", Toast.LENGTH_LONG).show();
```

```java
21      }
22      public int onStartCommand(Intent intent, int flags, int startId)
23      {
24          super.onStartCommand(intent, flags, startId);
25          play.start();         ← 开始播放音乐
26          Toast.makeText(this, "启动后台服务程序,播放音乐...",
27                  Toast.LENGTH_LONG).show();
28          return START_STICKY;
29      }
30      public void onDestroy()
31      {
32        play.release();
33        super.onDestroy();
34        Toast.makeText(this, "销毁后台服务!", Toast.LENGTH_LONG).show();
35      }
36  }
```

（3）启动后台服务的主控程序 MainActivity.java。

```java
1   package com.example.ex5_1;
2   import androidx.appcompat.app.AppCompatActivity;
3   import android.os.Bundle;
4   import android.content.Intent;
5   import android.view.View;
6   import android.view.View.OnClickListener;
7   import android.widget.Button;
8   import android.widget.TextView;
9   public class MainActivity extends Activity
10  {
11      Button startbtn, stopbtn;
12      Intent intent;
13      static TextView txt;
14      @Override
15      public void onCreate(Bundle savedInstanceState)
16      {
17          super.onCreate(savedInstanceState);
18          setContentView(R.layout.main);
19          startbtn=(Button)findViewById(R.id.butn1);
20          stopbtn=(Button)findViewById(R.id.butn2);
21          startbtn.setOnClickListener(new mClick());
22          stopbtn.setOnClickListener(new mClick());
23          txt=(TextView)findViewById(R.id.text1);
24          intent = new Intent(MainActivity.this, AudioSrv.class);   ← 创建 Intent 对象
25      }
26      class mClick implements OnClickListener    //定义一个类实现监听接口
27      {
28          public void onClick(View v)
```

```
29        {
30           if(v == startbtn)
31           {
32             MainActivity.this.startService(intent);   ← 启动 Intent 关联的 Service
33             txt.setText("start service .......");
34           }
35           else if(v == stopbtn)
36           {
37              MainActivity.this.stopService(intent);   ← 终止后台服务
38           }
39        }
40     }
41 }
```

（4）修改配置文件 AndroidManifest.xml。

在配置文件 AndroidManifest.xml 的<application>标签中添加以下代码：

```
<service android:enabled="true" android:name=".AudioSrv" />
```

修改后完整的 AndroidManifest.xml 文件如下：

```
1  <?xml version="1.0" encoding="utf-8"?>
2  <manifest xmlns:android="http://schemas.android.com/apk/res/android"
3      package="com.example.ex5_1" />
4      <application
5         android:allowBackup="true"
6         android:icon="@mipmap/ic_launcher"
7         android:label="ex5_1 "
8         android:roundIcon="@mipmap/ic_launcher_round"
9         android:supportsRtl="true"
10        android:theme="@style/Theme.ex5_1">
11        <activity
12           android:name=".MainActivity" >
13             <intent-filter>
14                <action android:name="android.intent.action.MAIN" />
15                <category android:name="android.intent.category.LAUNCHER" />
16             </intent-filter>
17        </activity>
18        <!-- 添加 AudioSrv 服务程序  -->
19        <service android:enabled="true" android:name=".AudioSrv" />
20     </application>
21 </manifest>
```

程序运行结果如图 5.1 所示。

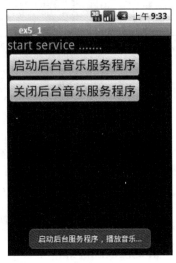

图 5.1　启动后台服务程序

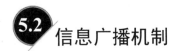 信息广播机制

Broadcast 是 Android 系统应用程序之间传递信息的一种机制。当系统之间需要传递某些信息时，不是通过诸如单击按钮之类的组件来触发事件，而是由系统自身通过系统调用来引发事件。这种系统调用是由 BroadcastReceiver 类实现的，把这种系统调用称为广播。BroadcastReceiver 是"广播接收者"的意思，顾名思义，它用来接收来自系统和应用中的广播信息。

在 Android 系统中有很多广播信息。例如，当开机时，系统会产生一条广播信息，接收到这条广播信息就能实现开机启动服务的功能；当网络状态改变时，系统会产生一条广播信息，接收到这条广播信息就能及时地做出提示和保存数据等操作；当电池电量改变时，系统会产生一条广播信息，接收到这条广播信息就能在电量低时告知用户及时保存进度等。

下面通过简单的应用示例来说明信息广播的设计方法。

【例 5-2】 一个简单的信息广播程序示例。

（1）设计主控文件 MainActivity.java。

```
1    package com.example.ex5_2;
2    import androidx.appcompat.app.AppCompatActivity;
3    import android.os.Bundle;
4    import android.content.Intent;
5    import android.view.View;
6    import android.widget.Button;
7    import android.widget.TextView;
8    public class MainActivity extends Activity
```

```
9   {
10      static TextView txt;        ← 要在类外调用,设置成静态变量
11      private  TestReceiver recevier;
12      private IntentFilter intentFilter;
13      @Override
14      public void onCreate(Bundle savedInstanceState)
15      {
16         super.onCreate(savedInstanceState);
17         setContentView(R.layout.main);
18         txt = (TextView)findViewById(R.id.txtView);
19         Button btn=(Button)findViewById(R.id.button);
20         btn.setOnClickListener(new mClick());
21      }

22      class mClick implements  View.OnClickListener
23      {
24       @Override
25       public void onClick(View v)
26       {
27         recevier = new TestReceiver();
28         intentFilter = new IntentFilter();    ← 创建过滤器
29         intentFilter.addAction("android.net.conn.CONNECTIVITY_CHANGE");
30         registerReceiver(recevier, intentFilter);   ← 注册广播
31         System.out.println("广播已经发送");
32       }
33      }
34   }
```

(2)设计广播接收器文件 TestReceiver.java。

```
1  package com.ex5_2;
2  import android.content.BroadcastReceiver;
3  import android.content.Context;
4  import android.content.Intent;
5
6  public class TestReceiver extends BroadcastReceiver   ← 定义广播接收器
7  {
8      @Override
9      public void onReceive(Context context, Intent intent)
10     {
11         String str = intent.getExtras().getString("hello");   ← 接收的广播数据
12         MainActivity.txt.setText(str);    ← 显示接收的数据
13     }
14 }
```

(3)设计配置文件 AndroidManifest.xml。

```
1   <?xml version="1.0" encoding="utf-8"?>
```

```
 2    <manifest xmlns:android="http://schemas.android.com/apk/res/android"
 3        package="com.example.ex5_2">

 4    <application
 5        android:allowBackup="true"
 6        android:icon="@mipmap/ic_launcher"
 7        android:label="@string/app_name"
 8        android:roundIcon="@mipmap/ic_launcher_round"
 9        android:supportsRtl="true"
10        android:theme="@style/Theme.ex5_2">
11        <activity android:name=".MainActivity">
12            <intent-filter>
13                <action android:name="android.intent.action.MAIN" />

14                <category android:name="android.intent.category.LAUNCHER" />
15            </intent-filter>
16        </activity>
17        <!-- 注册对应的广播接收类 -->
18        <receiver  android:name=".TestReceiver">      新增语句
19        </receiver>
20    </application>
21    </manifest>
```

程序运行结果如图 5.2 所示。

图 5.2 简单的广播示例

【例 5-3】 由一个后台服务广播音乐的播放、暂停或停止信息,接收器接收到信息后,执行改变用户界面按钮上文本的操作。

在本例中创建了 3 个类,即 MainActivity、AudioService 和 Broadcast,MainActivity 负责用户的交互界面,并启动后台服务;AudioService 是 Service 的子类,在后台提供播放、暂停或停止音乐等功能,同时发送改变交互界面的广播信息;Broadcast 是 BroadcastReceiver 的子类,负责接收广播信息,更改交互界面。

(1) 设计界面布局。

在界面布局中,设置一个文本标签和 3 个按钮,如图 5.3 所示。

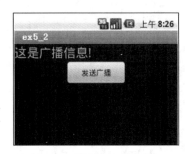

图 5.3 后台服务广播音乐的界面布局设计

（2）设计主控文件 MainActivity.java。

```
1   package com.example.ex5_3;
2   import androidx.appcompat.app.AppCompatActivity;
3   import android.os.Bundle;
4   import android.content.Intent;
5   import android.view.View;
6   import android.widget.Button;
7   import android.widget.TextView;

8   public class MainActivity extends AppCompatActivity
9   {
10      static Button startbtn, stopbtn, paubtn;
11      static TextView txt;
12      Intent intent;

13      @Override
14      public void onCreate(Bundle savedInstanceState)
15      {
16          super.onCreate(savedInstanceState);
17          setContentView(R.layout.main);
18          startbtn=(Button)findViewById(R.id.button);
19          paubtn = (Button)findViewById(R.id.button2);
20          stopbtn=(Button)findViewById(R.id.button3);
21          txt=(TextView)findViewById(R.id.textView);
22          intent = new Intent(MainActivity.this, AudioSrv.class);
23          startbtn.setOnClickListener(new mClick());
24          paubtn.setOnClickListener(new mClick());
25          stopbtn.setOnClickListener(new mClick());
26      }

27      class mClick implements View.OnClickListener   //定义实现监听接口类
28      {
29          public void onClick(View v)
30          {
31              if(v == startbtn)
32              {
33                  intent.putExtra("play",1);
34                  MainActivity.this.startService(intent);//开启服务
35                  MainActivity.this.startService(intent);
36                  System.out.println("(1)启动音乐");
37              }
38              else if(v == paubtn)
39              {
40                  intent.putExtra("play",2);
41                  MainActivity.this.startService(intent);
42                  System.out.println("(2)暂停音乐");
```

```
43              }
44            else if(v == stopbtn)
45            {
46                intent.putExtra("play",3);
47                MainActivity.this.startService(intent);
48                System.out.println("(3)停止音乐");
49            }
50        }
51    }
52 }
```

(3) 设计后台服务程序 AudioService.java。

```
1   package com.example.ex5_3;
2   import android.app.Service;
3   import android.content.Intent;
4   import android.content.IntentFilter;
5   import android.media.MediaPlayer;
6   import android.os.IBinder;
7   import android.widget.Toast;

8   public class  AudioService extends Service
9   {
10    public MediaPlayer play;
11    private AudioReceiver receiver;   //接收广播类
12    private IntentFilter intentFilter;
13    public IBinder onBind(Intent intent)
14    {
15       return null;
16    }
17    public void onCreate() //创建后台服务
18    {
19      super.onCreate();
20      play = MediaPlayer.create(this, R.raw.mtest);
21      Toast.makeText(this, "创建后台服务...", Toast.LENGTH_LONG).show();
22    }
23    void on_Start(){    //创建广播
24       receiver = new AudioReceiver();
25       intentFilter = new IntentFilter();
26       intentFilter.addAction("android.net.conn.CONNECTIVITY_CHANGE");
27       registerReceiver(receiver, intentFilter);
28    }
29    public int onStartCommand(Intent intent, int flags, int startId)
30    {
31      super.onStartCommand(intent, flags, startId);
32      switch (intent.getIntExtra("play",-1)){
33        case 1:
```

```
34       play.start();    //开始
35       on_Start();      //启动广播
36       Toast.makeText(this,"启动后台服务,播放音乐...",
37                  Toast.LENGTH_SHORT).show();//提示
38       break;
39    case 2:
40       if (play!=null&&play.isPlaying()){
41          play.pause();//暂停
42          Toast.makeText(this,"暂停...",Toast.LENGTH_SHORT).show();
43       }
44       else {
45          play.start();//继续
46          Toast.makeText(this,"继续...",Toast.LENGTH_SHORT).show();
47       }
48       break;
49    case 3:
50       play.stop();//停止
51       play.release();//释放内存
52       Toast.makeText(this,"销毁后台服务...",Toast.LENGTH_SHORT).show();
53       break;
54    }
55    return START_STICKY;
56  }
57  public void onDestroy()
58  {
59     play.release();
60     super.onDestroy();
61     Toast.makeText(this, "销毁后台服务!", Toast.LENGTH_LONG).show();
62  }
63 }
```

(4) 设计广播接收器程序 Broadcast.java。

```
1  package com.example.ex5_3;
2  import android.content.BroadcastReceiver;
3  import android.content.Context;
4  import android.content.Intent;
5  /* 广播接收器 */
6  class Broadcast extends  BroadcastReceiver
7  {
8     @Override
9     public void onReceive(Context context, Intent intent)
10    {
11        String action = intent.getAction();
12        if(action != null) {
13           String str = "正在播放音乐";
14           System.out.println(str);
```

```
15              MainActivity.startbtn.setText(str);
16              MainActivity.txt.setText(action);
17          }
18      }
19  }
```

(5) 在配置文件 AndroidManifest.xml 中注册后台服务和广播。

```
1   <?xml version="1.0" encoding="utf-8"?>
2   <manifest xmlns:android="http://schemas.android.com/apk/res/android"
3     package="com.example.ex5_3">
4     <application
5        android:allowBackup="true"
6        android:icon="@mipmap/ic_launcher"
7        android:label="@string/app_name"
8        android:roundIcon="@mipmap/ic_launcher_round"
9        android:supportsRtl="true"
10       android:theme="@style/Theme.ex5_3">
11       <activity android:name=".MainActivity">
12           <intent-filter>
13              <action android:name="android.intent.action.MAIN" />
14              <category android:name="android.intent.category.LAUNCHER" />
15           </intent-filter>
16       </activity>
17       <!-- 注册后台服务类 -->
18       <service android:enabled="true"  android:name=".AudioSrv"/>
19       <!-- 注册广播接收类 -->
20       <receiver android:name=".AudioReceiver"></receiver>
21    </application>
22  </manifest>
```

程序运行结果如图 5.4 所示。

图 5.4 运行后台广播服务

5.3 系统服务

在第 4 章中学习了应用 Intent 进行参数传递，Android 系统提供了很多标准的系统服务，这些系统服务都可以很简单地通过 Intent 进行广播。

5.3.1 Android 的系统服务

Android 系统提供了大量的标准系统服务，这些系统服务用于完成不同的功能，Android 的系统服务如表 5-2 所示。

表 5-2 Android 的系统服务

系 统 服 务	作 用
WINDOW_SERVICE ("window")	窗体管理服务
LAYOUT_INFLATER_SERVICE ("layout_inflater")	布局管理服务
ACTIVITY_SERVICE ("activity")	Activity 管理服务
POWER_SERVICE ("power")	电源管理服务
ALARM_SERVICE ("alarm")	时钟管理服务
NOTIFICATION_SERVICE ("notification")	通知管理服务
KEYGUARD_SERVICE ("keyguard")	键盘锁服务
LOCATION_SERVICE ("location")	基于地图的位置服务
SEARCH_SERVICE ("search")	搜索服务
VIBRATOR_SERVICE ("vibrator")	振动管理服务
CONNECTIVITY_SERVICE ("connection")	网络连接服务
WIFI_SERVICE ("wifi")	Wi-Fi 连接服务
INPUT_METHOD_SERVICE ("input_method")	输入法管理服务
TELEPHONY_SERVICE ("telephony")	电话服务
DOWNLOAD_SERVICE ("download")	HTTP 协议的下载服务

系统服务实际上可以看作是一个对象，通过 Activity 类的 getSystemService 方法可以获得指定的对象（系统服务）。下面详细讲解几个常见的系统服务。

5.3.2 系统通知服务

Notification 是 Android 系统的一种通知服务，当手机来电、来短信、闹钟铃声响时，在状态栏显示通知的图标和文字，提示用户处理，拖动状态栏可以查看这些信息。

使用 Notification 时需要配置一些属性，表 5-3 列出了 Notification 的部分常见属性。

系统通知服务 Notification 由系统通知管理对象 NotificationManager 进行管理及发布通知，由 getSystemService(NOTIFICATION_SERVICE)创建 NotificationManager 对象。

表 5-3　Notification 的部分常见属性

属　性	说　　明
setContentTitle	设置通知的标题（位于第 1 行）
setContentText	设置通知的文本（位于第 2 行）
setSmallIcon	通知布局中使用的小图标
setContentIntent	在点击通知时发送的内容
setLargeIcon	设置显示在通知中的大图标
setAutoCancel	在通知栏点击通知是否自动取消显示

```
NotificationManager n_Manager =
    (NotificationManager)getSystemService(NOTIFICATION_SERVICE);
```

NotificationManager 对象通过 notify(int id, Notification notification) 方法把通知发送到状态栏，通过 cancelAll() 方法取消以前显示的所有通知。

【例 5-4】 在状态栏中显示系统通知服务的应用示例。

在界面布局文件中设置两个按钮，分别为"发送系统通知"和"删除通知"。设计控制文件 MainActivity.java 如下：

```
1    package com.example.ex5_4;

2    import androidx.appcompat.app.AppCompatActivity;
3    import android.os.Bundle;
4    import androidx.core.app.NotificationCompat;
5    import android.app.NotificationChannel;
6    import android.content.Context;
7    import android.graphics.BitmapFactory;
8    import android.os.Build;
9    import android.app.Notification;
10   import android.app.NotificationManager;
11   import android.app.PendingIntent;
12   import android.content.Intent;
13   import android.view.View;
14   import android.widget.Button;

15   public class MainActivity extends AppCompatActivity {
16       NotificationManager n_Manager;
17       Notification notification;
18       Button btn1, btn2;
19       private final static int NOTIFY_ID = 100;

20       @Override
21       protected void onCreate(Bundle savedInstanceState) {
22           super.onCreate(savedInstanceState);
23           setContentView(R.layout.activity_main);
24           btn1=(Button)findViewById(R.id.btn1);
```

```
25       btn1.setOnClickListener(new mClick());
26       btn2=(Button)findViewById(R.id.btn2);
27       btn2.setOnClickListener(new mClick());
28    }
29    class mClick implements View.OnClickListener{
30      @Override
31      public void onClick(View v) {
32       String service = Context.NOTIFICATION_SERVICE;
33       n_Manager= (NotificationManager) MainActivity.this
34                    .getSystemService(service);
35       if(v==btn1)
36       {
37        showNotification();
38       }
39       else if(v==btn2)
40       {
41          n_Manager.cancelAll();
42       }
43     }
44     private void  showNotification(){
45      int icon1 = R.drawable.ic_con;     //事先准备好图标文件
46      int icon2 = R.drawable.icon_tz;
47      CharSequence tickerText = "紧急通知,程序已启动";
48      long when = System.currentTimeMillis();
49      String  CHANNEL_ID = "zsm_id";    //应用频道id唯一值
50      String CHANNEL_NAME = "zsm_name";
51      Intent intent =new Intent(MainActivity.this,MainActivity.class);
52      PendingIntent pintent = PendingIntent.getActivity(
53          MainActivity.this,
54          1001,
55          intent,
56          PendingIntent.FLAG_UPDATE_CURRENT);
57       notification = new NotificationCompat.Builder(
58          MainActivity.this, CHANNEL_ID)
59          .setSmallIcon(icon2)            //设置小图标
60          .setContentTitle("通知")
61          .setContentText(tickerText)     //通知内容
62          .setContentIntent(pintent)
63          .setLargeIcon(BitmapFactory
64          .decodeResource(getResources(), icon1))
65          .build();
66     //Android 8.0 以上需添加渠道
67      if (Build.VERSION.SDK_INT >= Build.VERSION_CODES.O) {
68         NotificationChannel notificationChannel =
69           new NotificationChannel(CHANNEL_ID,
70           CHANNEL_NAME, NotificationManager.IMPORTANCE_LOW);
71        n_Manager.createNotificationChannel(notificationChannel);
```

```
72          }
73          n_Manager.notify(NOTIFY_ID, notification);
74      }
75  }
76 }
```

程序运行结果如图 5.5 所示。

(a) 单击"发送系统通知"按钮，显示小图标　　(b) 向下拖动状态栏，显示通知内容

图 5.5　在状态栏中显示系统通知

习题 5

1. 结合例 5-1 和例 5-3，编写一个具有较完善功能的后台音乐播放器。
2. 编写一个短信服务平台。

第6章 网络通信

6.1 Web 视图

6.1.1 浏览器引擎

WebKit 是一个开源的浏览器引擎。Webkit 内核具有非常好的网页解析机制,很多应用系统都使用 WebKit 做浏览器的内核。例如,Google 的 Android 系统、Apple 的 iOS 系统、Nokia 的 Series 60 browser 系统所使用的 Browser 内核引擎都是基于 Webkit 的。Webkit 所包含的 WebCore 排版引擎和 JSCore 引擎来自于 KDE 的 KHTML 和 KJS,它们拥有清晰的源码结构和极快的渲染速度。

Android 对 WebKit 做了进一步的封装,并提供了丰富的 API。Android 平台的 WebKit 模块由 Java 层和 WebKit 库两个部分组成,Java 层负责与 Android 应用程序进行通信,而 WebKit 类库负责实际的网页排版处理。WebKit 包中的几个重要类如表 6-1 所示。

表 6-1 WebKit 包中的几个重要类

类 名	说 明
WebSettings	用于设置 WebView 的特征、属性等
WebView	显示 Web 页面的视图对象,用于网页数据载入、显示等操作
WebViewClient	在 Web 视图中帮助处理各种通知、请求事件
WebChromeClient	Google 浏览器 Chrome 的基类,辅助 WebView 处理 JavaScript 对话框、网站的标题、网站的图标、加载进度条等

6.1.2 Web 视图对象

1. WebView 类

在 WebKit 的 API 包中,最重要、最常用的类是 Android.Webkit.WebView。WebView 类是 WebKit 模块 Java 层的视图类,所有需要使用 Web 浏览功能的 Android 应用程序都要创建该视图对象,用于显示和处理请求的网络资源。目前,WebKit 模块支持 HTTP、HTTPS、

FTP 以及 JavaScript 请求。WebView 作为应用程序的 UI 接口，为用户提供了一系列的网页浏览、用户交互接口，客户程序通过这些接口访问 WebKit 核心代码。

WebView 类的常用方法如表 6-2 所示。

表 6-2 WebView 类的常用方法

方　　法	说　　明
WebView(Context context)	构造方法
loadUrl(String url)	加载 URL 网站页面
loadData(String data, String mimeType, String encod)	显示 HTML 格式的 Web 视图
reload()	重新加载网页
getSettings()	获取 WebSettings 对象
goBack()	返回上一页面
goForward()	前往下一页面
clearHistory()	清除历史记录
addJavascriptInterface (Object obj, String interfaceName)	将对象绑定到 Javascript，允许从网页控制 Android 程序、调用该对象的方法

2. 使用 WebView 的说明

（1）设置 WebView 基本信息。

- 如果访问的页面中有 Javascript，则 WebView 必须设置支持 JavaScript：

```
webview.getSettings().setJavaScriptEnabled(true);
```

- 触摸焦点起作用：

```
requestFocus();
```

- 取消滚动条：

```
this.setScrollBarStyle(SCROLLBARS_OUTSIDE_OVERLAY);
```

（2）设置 WebView 要显示的网页。

- 互联网用 webView.loadUrl("http://www.google.com");
- 本地文件用 webView.loadUrl("file:///android_asset/XX.html")，本地文件要存放在项目的 assets 目录中。

（3）用 WebView 点击链接看了很多页面以后，如果不做任何处理，按 BackSpace 键浏览器会调用 finish()结束自身的运行；如果希望浏览的网页回退而不是退出浏览器，需要在当前 Activity 中覆盖 Activity 类的 onKeyDown(int keyCoder, KeyEvent event)方法处理该 Back 事件。

```
public boolean onKeyDown(int keyCoder,KeyEvent event)
{
    if(webView.canGoBack() && keyCoder == KeyEvent.KEYCODE_BACK)
    {
```

```
            webview.goBack();      //goBack()表示返回 WebView 的上一页面
            return true;
        }
        return false;
    }
```

【例 6-1】 应用 WebView 对象浏览网页。

(1) 设计界面布局文件 activity_main.xml。

在界面布局中,设置了一个文本编辑框,用于输入网址;设置了一个按钮,用于打开网页;还设置了一个网页视图组件 WebView,用于显示网页。

(2) 设计控制文件 MainActivity.java。

```
1   package com.example.ex6_1;
2   import androidx.appcompat.app.AppCompatActivity;
3   import android.os.Bundle;
4   import android.view.View;
5   import android.view.View.OnClickListener;
6   import android.webkit.WebView;
7   import android.widget.Button;
8   import android.widget.EditText;
9   public class MainActivity extends AppCompatActivity
10  {
11      WebView webView;
12      Button openWebBtn;
13      EditText editText;
14      @Override
15      public void onCreate(Bundle savedInstanceState)
16      {
17        super.onCreate(savedInstanceState);
18        setContentView(R.layout.activity_main);
19        openWebBtn = (Button)findViewById(R.id.button);
20        editText =(EditText)findViewById(R.id.editText);
21        webView = (WebView)findViewById(R.id.webView);
22        openWebBtn.setOnClickListener(new mClick());
23      }
24      class mClick implements OnClickListener
25      {
26         public void onClick(View arg0)
27         {
28              String url = editText.getText().toString();
29              webView.loadUrl("http://" + url);
30         }
31      }
32  }
```

(3) 在配置文件中加入网络权限。网络程序需要在配置文件 AndroidManifest.xml 中加入允许访问网络的权限语句:

```
<uses-permission android:name="android.permission.INTERNET" />
```

另外,为了允许使用 HTTP 的网站(默认仅支持 HTTPS),再增加<application>的属性语句:

```
android:usesCleartextTraffic="true"
```

添加权限后的 AndroidManifest.xml 程序如下:

```
1   <?xml version="1.0" encoding="utf-8"?>
2   <manifest xmlns:android="http://schemas.android.com/apk/res/android"
3       package="com.example.ex6_1">

4       <application
5           android:allowBackup="true"
6           android:icon="@mipmap/ic_launcher"
7           android:label="@string/app_name"
8           android:roundIcon="@mipmap/ic_launcher_round"
9           android:supportsRtl="true"
10          android:usesCleartextTraffic="true"
11          android:theme="@style/Theme.ex6_1">
12          <activity android:name=".MainActivity">
13              <intent-filter>
14                  <action android:name="android.intent.action.MAIN" />
15                  <category android:name="android.intent.category.LAUNCHER" />
16              </intent-filter>
17          </activity>
18      </application>
19      <uses-permission android:name="android.permission.INTERNET" />
20  </manifest>
```

程序运行结果如图 6.1 所示。

图 6.1 用 WebView 显示网页

6.1.3 调用 JavaScript

1. 几个辅助类

（1）Websettings 类。

WebView 对象刚创建时使用的是系统的默认设置，当需要对 WebView 对象的属性等做自定义设置时，需要用到 Websettings 类。Websettings 类的常用方法如表 6-3 所示。

表 6-3　Websettings 类的常用方法

方　　法	说　　明
setAllowFileAccess(boolean flag)	设置是否允许访问文件数据
setJavaScriptEnabled(boolean flag)	设置是否支持 JavaScript 脚本
setBuiltInZoomControls(boolean flag)	设置是否支持缩放
setBlockNetworkImage(boolean flag)	设置是否禁止显示图片，true 为禁止显示
setDefaultFontSize(int size)	设置默认字体大小，取值为 1~72
setTextZoom(int textZoom)	设置页面文字缩放的百分比，默认为 100

（2）WebViewClient 类。

WebViewClient 类用于对 WebView 对象中各种事件的处理，通过重写提供的这些事件方法，可以对 WebView 对象在页面载入、资源载入、页面访问错误等情况发生时进行各种操作。 WebViewClient 类的常用方法如表 6-4 所示。

表 6-4　WebViewClient 类的常用方法

方　　法	说　　明
onLoadResource(WebView view, String url)	通知 WebView 加载 url 指定的资源时触发
onPageStarted(WebView view, String url, Bitmap favicon)	页面开始加载时触发
onPageFinished(WebView view, String url)	页面加载完毕时触发

（3）WebChromeClient 类。

WebChromeClient 是辅助 WebView 处理 JavaScript 对话框、网站的标题、网站的图标、加载进度条等操作的类，其常用方法如表 6-5 所示。

表 6-5　WebChromeClient 类的常用方法

方　　法	说　　明
onJsAlert(WebView view, String url, String message, JsResult result)	处理 JavaScript 的 Alert 对话框
onJsPrompt(WebView view, String url, String message, String defaultValue, JsPromptResult result)	处理 JavaScript 的 Prompt 提示对话框
onCloseWindow(WebView window)	关闭 WebView

2. 调用本地 HTML 网页文件的 JavaScript

用户可以在 Android 程序中调用本地的 HTML 网页文件的 JavaScript，如下面例子所示。

【例 6-2】 在 Android 程序中调用本地的 HTML 程序示例。

（1）在 Android Studio 编辑器中，首先调整成 Project 模式，然后在 main 目录下新建 assets 目录，在 assets 目录下新建一个 HTML 程序 test.html。通常，assets 目录存放应用程序所使用的外部资源文件，而 res 目录存放应用程序自身的资源文件，如图 6.2 所示。

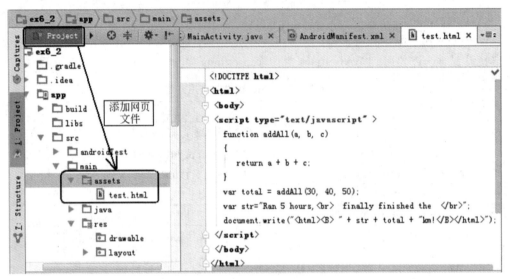

图 6.2 在 assets 目录下新建 test.html

HTML 程序 test.html 的代码如下：

```
1    <HTML>
2    <head>
3      <title> 一个简单的 Javascript 示例 </title>
4    </head>
5    <body>
6      <script language="javascript" type="text/javascript" >
7        function addAll(a, b, c)
8        {
9          return a + b + c;
10       }
11       var total = addAll(30, 40, 50);
12       var str="Ran 5 hours,<br> finally finished the "<br>;
13       document.write("<html><B> " + str + total + " km!</B></html>");
14     </script>
15   </body>
16   </HTML>
```

（2）设计界面布局文件。

在界面布局中，设置一个网页视图组件 WebView，用于显示网页。

（3）设计控制文件。

```
1    package com.example.ex6_2;
2    import androidx.appcompat.app.AppCompatActivity;
```

```
3   import android.os.Bundle;
4   import android.webkit.WebChromeClient;
5   import android.webkit.WebSettings;
6   import android.webkit.WebView;
7   public class MainActivity extends Activity
8   {
9     WebView webView;
10    @Override
11    public void onCreate(Bundle savedInstanceState)
12    {
13      super.onCreate(savedInstanceState);
14      setContentView(R.layout.activity_main);
15      webView = (WebView) findViewById(R.id.webView1);
16      WebSettings webSettings = webView.getSettings();
17      webSettings.setAllowFileAccess(true);// 设置允许访问文件数据
18      webSettings.setJavaScriptEnabled(true);// 设置支持 JavaScript 脚本
19      webSettings.setBuiltInZoomControls(true);// 设置支持缩放
20      webSettings.setDefaultFontSize (24);
21      webView.setWebChromeClient(mWebChromeClient);
22      webView.loadUrl("file:///android_asset/test1.html");
23    }
24  }
```

程序运行结果如图 6.3 所示。

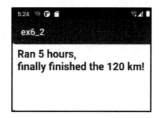

图 6.3　调用 HTML 的 JavaScript

【例 6-3】 用 Android 程序操作 JavaScript 对话框。

（1）在 Android Studio 编辑器中，首先调整成 Project 模式，然后在 main 目录下新建 assets 目录，在 assets 目录下新建一个 JavaScript 对话框程序 test1.html，其代码如下：

```
1   <html>
2   <head>
3   <title>javascript 与 android 交互</title>
4   </head>
5   <script type="text/javascript">
6     function show_alert()
7     {
8       var a = document.getElementById("text").value;
9       alert("Hello " + a );
10    }
```

JavaScript 对话框

```
11    </script>
12    <body>
13    <form action="">
14        <input type="text" id="text" value=""/>
15        <input type="button" id="button"
16            onclick="window.test.android_show()"
17            value="call Android"/>
18    </form>
19    </body>
20    </html>
```

调用 Android 程序标记为 test 的实例对象的函数

（2）设计界面布局文件。

```
1   <?xml version="1.0" encoding="utf-8"?>
2   <androidx.constraintlayout.widget.ConstraintLayout
3    xmlns:android="http://schemas.android.com/apk/res/android"
4      xmlns:app="http://schemas.android.com/apk/res-auto"
5      xmlns:tools="http://schemas.android.com/tools"
6      android:layout_width="match_parent"
7      android:layout_height="match_parent"
8      tools:context=".MainActivity">

9      <WebView
10         android:id="@+id/webView"
11         android:layout_width="0dp"
12         android:layout_height="0dp" />
13  </androidx.constraintlayout.widget.ConstraintLayout>
```

（3）设计控制文件。

```
1   package com.example.ex6_3;
2   import androidx.appcompat.app.AppCompatActivity;
3   import android.os.Bundle;
4   import android.os.Handler;
5   import android.webkit.JavascriptInterface;
6   import android.webkit.JsResult;
7   import android.webkit.WebChromeClient;
8   import android.webkit.WebSettings;
9   import android.webkit.WebView;
10  import android.widget.Toast;

11  public class MainActivity extends AppCompatActivity
12  {
13      WebView webView;
14      Handler handler = new Handler();
15      MWebChromeClient mWebChromeClient;
16      @Override
17      public void onCreate(Bundle savedInstanceState)
```

```
18      {
19          super.onCreate(savedInstanceState);
20          setContentView(R.layout.activity_main);
21          webView = (WebView) findViewById(R.id.webView1);
22          WebSettings webSettings = webView.getSettings();
23          webSettings.setAllowFileAccess(true);// 设置允许访问文件数据
24          webSettings.setJavaScriptEnabled(true);// 设置支持 JavaScript 脚本
25          webSettings.setBuiltInZoomControls(true);// 设置支持缩放
26          webSettings.setDefaultFontSize (24);
27          MObject mObject = new MObject();
28          webView.addJavascriptInterface(mObject, "test");
29          mWebChromeClient = new MWebChromeClient();
30          webView.setWebChromeClient(mWebChromeClient);
31          webView.loadUrl("file:///android_asset/test1.html");
32      }
33      class MObject extends Object
34      {
35          @JavascriptInterface         ◄── 为了安全问题,"@JavascriptInterface"注解不可少
36          public void  android_show()
37          {
38            handler.post(new Runnable()
39            {
40                public void run()
41                {
42                   System.out.println("提示:调用了多线程的 run()方法!!");
43                   webView.loadUrl("javascript:show_alert()");  ◄── 调用 javascript 函数
44                }
45           });
46         }
47      }
48      class MWebChromeClient extends WebChromeClient
49      {
50        @Override
51        public boolean onJsAlert(WebView view,    ◄── 处理 JavaScript 的 Alert 对话框
52                  String url,String message,JsResult result)
53        {
54           Toast.makeText(getApplicationContext(), message,
55                   Toast.LENGTH_LONG).show();
56           return true;
57        }
58      }
59 }
```

程序运行结果如图 6.4 所示。

图 6.4　操作 JavaScript 对话框

6.2　基于 TCP 的网络程序设计

本节将介绍 Android 编写网络通信程序的一些实例，其中重点介绍客户端 / 服务器的应用程序及 Web 视图应用程序的设计方法。

■ 6.2.1　网络编程的基础知识

1. IP 地址

在网络中连接了很多计算机，假设计算机 A 向计算机 B 发送信息，若网络中还有第 3 台计算机 C，那么主机 A 怎么知道信息被正确地传送到主机 B 而不是被传送到主机 C 了呢？如图 6.5 所示。

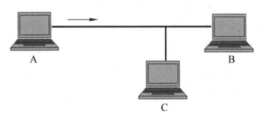

图 6.5　主机 A 向主机 B 发送信息

网络中的每台计算机都必须有一个唯一的 IP 地址作为标识，这个数通常写作一组由"."分隔的十进制数。IP 地址均由 4 个部分组成，每部分的范围都是 0～255，以表示 8 位地址。

值得注意的是，IP 地址都是 32 位地址，这是 IP 版本 4（简称 IPv4）规定的。目前由于 IPv4 地址已近耗尽，所以 IPv6 地址正逐渐代替 IPv4 地址，IPv6 地址则是 128 位无符号整数。

在 Java.net 包中，IP 地址由一个称作 InetAddress 的特殊类来描述。这个类提供了 3 个用来获得一个 InetAddress 类的实例的静态方法。这 3 个方法如下。

- getLocalHost( )：返回一个本地主机的 IP 地址。
- getByName(String host)：返回对应于指定主机的 IP 地址。
- getAllByName(String host)：对于某个主机有多个 IP 地址（多宿主机），可用于得到一个 IP 地址数组。

此外，对一个 InetAddress 的实例可以使用以下方法。

- getAddress( )：获得一个用字节数组形式表示的 IP 地址。
- getHostName( )：进行反向查询，获得对应于某个 IP 地址的主机名。

【例 6-4】 通过域名查找 IP 地址。

```
1   package com.example.ex6_4;
2   import androidx.appcompat.app.AppCompatActivity;
3   import android.os.Bundle;
4   import android.view.View;
5   import android.view.View.OnClickListener;
6   import android.widget.Button;
7   import android.widget.Toast;
8   import java.net.InetAddress;
9   import java.net.NetworkInterface;
10  import java.net.SocketException;
11  import java.util.ArrayList;
12  import java.util.Collections;
13  import java.net.Inet4Address;


14  public class MainActivity extends AppCompatActivity
15  {
16      Button IPBtn;
17      String str;
18      @Override
19      public void onCreate(Bundle savedInstanceState)
20      {
21          super.onCreate(savedInstanceState);
22          setContentView(R.layout.activity_main);
23          IPBtn=(Button)findViewById(R.id.button1);
24          IPBtn.setOnClickListener(new mClick());
25      }

26      class mClick implements View.OnClickListener
27      {
28          @Override
29          public void onClick(View v)
30          {
31              try{
32                  str = getLocalIPAddress();;
33              } catch(Exception e1)
34              {
35                  e1.printStackTrace();
36              }
37              Toast.makeText(MainActivity.this, str, Toast.LENGTH_LONG).show();
38          }
39      }
```

```
40  public String getLocalIPAddress()
41  {
42      try {
43          String ipv4;
44          ArrayList<NetworkInterface> nilist =
45          Collections.list(NetworkInterface.getNetworkInterfaces());
46          for (NetworkInterface ni: nilist)
47          {
48              ArrayList<InetAddress> ialist =
49                          Collections.list(ni.getInetAddresses());
50              for (InetAddress address: ialist)
51              {
52                  ipv4 = address.getHostAddress();
53                  if (!address.isLoopbackAddress()
54                      && address instanceof Inet4Address)
55                  {
56                      return ipv4;
57                  }
58              }
59          }
60      } catch (SocketException ex) {
61          Log.e("localip", ex.toString());
62      }
63      return null;
64  }
65  }
```

网络程序需要在配置文件 AndroidManifest.xml 中加入允许访问网络的权限语句：

```
<uses-permission android:name="android.permission.INTERNET" />
<uses-permission android:name="android.permission.ACCESS_NETWORK_STATE" />
```

程序运行结果如图 6.6 所示。

图 6.6　通过域名查找 IP 地址

2. 端口

由于一台计算机上可同时运行多个网络程序，IP 地址只能保证把数据信息送到该计算机，但无法知道要把这些数据交给该主机上的哪个网络程序，因此用"端口号"来标识正在计算机上运行的进程（程序）。每个被发送的网络数据包也都包含有"端口号"，用于将该数据帧交给具有相同端口号的应用程序来处理。

例如，在一个网络程序指定了自己所用的端口号为 52000，那么其他网络程序（比如端口号为 13）发送给这个网络程序的数据包必须包含 52000 端口号，当数据到达计算机后，驱动程序根据数据包中的端口号就知道要将这个数据包交给这个网络程序，如图 6.7 所示。

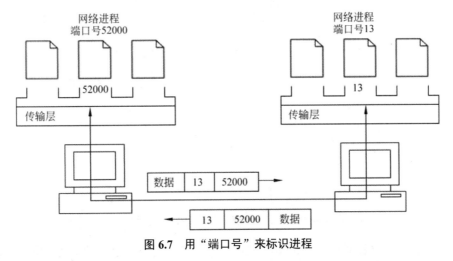

图 6.7 用"端口号"来标识进程

端口号是一个整数，其取值范围为 0～65535。同一台计算机上不能同时运行两个有相同端口号的进程。通常 0～1023 的端口号作为保留端口号，用于一些网络系统服务和应用，用户的普通网络应用程序应该使用 1024 以后的端口号，从而避免端口号冲突。

3. TCP 与 UDP

在网络协议中，有两个高级协议是网络应用程序编写时常用的，它们是传输控制协议（Transmission Control Protocol，TCP）和用户数据报协议（User Datagram Protocol，UDP）。

TCP 是面向连接的通信协议，TCP 提供两台计算机之间的可靠、无差错的数据传输。在应用程序利用 TCP 进行通信时，信息源与信息目标之间会建立一个虚连接。这个连接一旦建立成功，两台计算机之间就可以把数据当作一个双向字节流进行交换。接收方对于接收到的每一个数据包都会发送一个确认信息，发送方只有收到接收方的确认信息后才发送下一个数据包。通过这种确认机制保证数据传输无差错。

UDP 是无连接通信协议，UDP 不保证可靠数据的传输。简单地说，如果一台主机向另一台主机发送数据，这一数据就会立即发送，而不管另一台主机是否已准备接收数据。如果另一台主机收到了数据，它不会确认收到与否。这一过程类似于从邮局发送信件，发信人无法确定收信人一定收到了发出去的信件。

4. 套接字

通过 IP 地址可以在网络上找到主机,通过端口号可以找到主机上正在运行的网络程序。在 TCP/IP 通信协议中,套接字（Socket）就是 IP 地址与端口号的组合。如图 6.8 所示,IP 地址 193.14.26.7 与端口号 13 组成一个套接字。

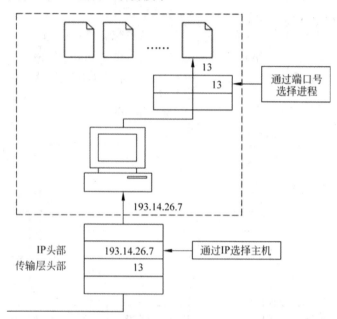

图 6.8　套接字是 IP 地址和端口号的组合

Java 使用了 TCP/IP 套接字机制,并使用一些类来实现套接字中的概念。Java 中的套接字提供了在一台处理机上执行的应用程序与在另一台处理机上执行的应用程序之间进行连接的功能。

准确地说,网络通信不能仅说成是两台计算机之间在通信,而是两台计算机上执行的网络应用程序（进程）之间在收发数据。

当两个网络程序需要通信时,它们可以通过使用 Socket 类建立套接字连接。我们可以把套接字连接想象为一个电话呼叫,当呼叫完成后,通话的任何一方都可以随时讲话。但是在最初建立呼叫时,必须有一方主动呼叫,而另一方正在监听铃声。这时,我们把呼叫方称为"客户端",把负责监听的一方称为"服务器端"。

5. 在客户端建立套接字 Socket 对象

在客户端使用 Socket 类,建立向指定服务器 IP 和端口号连接的套接字,其构造方法如下:

Socket(host_IP, port);

其中,host_IP 是服务器的 IP 地址,port 是一个端口号。

由于建立 Socket 对象可能发生 IOException 异常,因此,在建立 Socket 对象时要使用 try-catch 结构处理异常事件。

Socket 的主要方法如下：

- getInputStream( )：获得一个输入流，读取从网络线路上传送来的数据信息。
- getOutputStream( )：获得一个输出流，将数据信息写入到网络线路上。

6．在服务器端建立套接字 Socket 对象

在编写 TCP 网络服务器程序时，首先要用 ServerSocket 类创建服务器 Socket，ServerSocket 类的构造方法如下：

```
ServerSocket(int port);
```

创建 ServerSocket 实例是不需要指定 IP 地址的，ServerSocket 总是处于监听本机端口的状态。

ServerSocket 类的主要方法如下：

```
Socket accept();
```

该方法用于在服务器端的指定端口监听客户机发起的连接请求，并与之连接，其返回值为 Socket 对象。

6.2.2 利用套接字 Socket 设计客户端/服务器系统程序

基于 TCP 的网络程序均采用客户端/服务器系统模式。利用套接字 Socket 设计客户端/服务器系统程序，并进行数据通信与传输，大致有以下几个步骤。

（1）在计算机上创建服务器端 ServerSocket，设置建立连接的端口号。
（2）创建 Android 客户端 Socket 对象，设置绑定的主机名称或 IP 地址，指定连接端口号。
（3）客户端 Socket 发起连接请求。
（4）建立连接。
（5）取得 InputStream 和 OutputStream。
（6）利用 InputStream 和 OutputStream 进行数据传输。
（7）关闭 Socket 和 ServerSocket。

客户端/服务器模式的连接请求与响应过程如图 6.9 所示。

【例 6-5】远程数据通信示例，本例由 Android 客户端程序和计算机上运行的服务器程序两部分组成。

（1）Android 客户端程序。

```
1   package com.example.ex6_5;
2   import android.appcompat.app.AppCompatActivity;
3   import android.os.Bundle;
4   import java.io.DataInputStream;
5   import java.io.DataOutputStream;
6   import java.net.Socket;
7   import android.os.StrictMode;
8   import android.view.View;
9   import android.widget.Button;
```

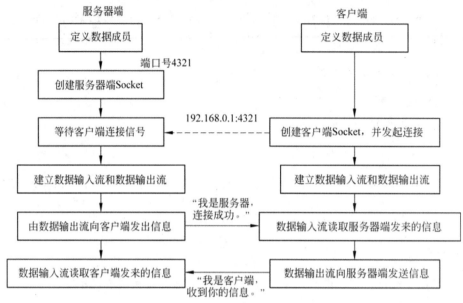

图 6.9 客户端/服务器模式

```
10    import android.widget.TextView;

11    public class SocketClientActivity extends AppCompatActivity
12    {
13      privateSocket socket=null;
14      privateDataInputStream dis=null;
15      privateDataOutputStream dos=null;
16      private TextView mTextView1;
17      private Button Button1;
18      String msg="";

19      @Override
20      public void onCreate(Bundle savedInstanceState)
21      {
22        super.onCreate(savedInstanceState);
23        setContentView(R.layout.activity_main);
24        mTextView1 = (TextView)findViewById(R.id.textView);
25        Button1 = (Button) findViewById(R.id.Button);
26        Button1.setOnClickListener(new mClick());
27        //以下代码避免程序出现NetworkOnMainThreadException 异常
28        StrictMode.setThreadPolicy(new  StrictMode
29                        .ThreadPolicy
30                        .Builder()
31                        .detectDiskWrites()
32                        .detectDiskReads()
33                        .detectNetwork()
34                        .penaltyLog()
35                        .build()  );
```

设置线程策略

```
36    }
37    class mClick implements View.OnClickListener {
38      @Override
39      public void onClick(View v)
40      {
41            Client();
42      }
43      public void Client()
44      {
45        try {
46        socket = new Socket("192.168.0.1", 4321);    ← 创建一个 Socket 流连接到目标
47        }catch (Exception ioe) {                        主机（请更换为目标主机的 IP）
48              System.out.print("socket  err ");
49        }
50        try {
51           // 创建输入流对象 dis 读取数据，创建输出流对象 dos 发送数据
52              dis = new DataInputStream(socket.getInputStream());
53              dos = new DataOutputStream(socket.getOutputStream());
54              dos.writeUTF("给我数据啊。。。");
55              dos.flush();
56        } catch (IOException ioe) {
57              System.out.print("DataStream create err ");
58            }
59        ReadStr();     ← 读取服务器端发来的数据
60        try{
61              Thread.sleep(500);
62              msg = "手机客户端发来贺电！";
63              WriteString(msg);    ← 发送数据
64              dis.close();
65              socket.close();
66        }catch (Exception ioe) {
67              System.out.println("socket close() err ......."); }
68        ReadStr();     ← 读取服务器端发来的第 2 条数据
69      }
70      // 写数据到 socket, 即发送数据
71      public void WriteString(String str)
72      {
73        try {
74              dos.writeUTF(str);    ← 发送字符串 str 的数据
75              dos.flush();
76              ReadStr();
77              socket.close();
78        } catch (IOException e) {
79              System.out.print("WriteString() err");
80        }
81      }
82      // 接收从 socket 返回的数据，即读取数据
```

```
83      public void ReadStr()
84      {
85         try {
86              if((msg =dis.readUTF())!=null)    ← 读取数据存放到字符串变量 msg
87              {
88                 mTextView1.append(msg);
89              }
90         } catch (IOException ioe) {
91              System.out.print("ReadStr() err ");
92         }
93      }
94  }
```

（2）服务器端程序。

```
1   import java.io.DataOutputStream;
2   import java.io.DataInputStream;
3   import java.io.IOException;
4   import java.net.ServerSocket;
5   import java.net.Socket;
6
7   public class Server
8   {
9     private ServerSocket ss;
10    private Socket socket;
11    private DataInputStream dis;
12    private DataOutputStream dos;
13    public Server()
14    {
15      new ServerThread().start();
16    }
17    class ServerThread extends Thread
18    {
19      public void run()
20      {
21        try {
22          ss=new ServerSocket(4321);          ← 实例化服务器端套接字对象
23          System.out.println("服务器启动了");
24          while(true)
25          {
26            socket=ss.accept();               ← 阻塞端口，等待客户端连接
27            System.out.println("有客户端连接到服务器");
28            dis =new DataInputStream(socket.getInputStream());
29            dos =new DataOutputStream(socket.getOutputStream());
30            dos.writeUTF("恭喜你，连接服务器成功！ \n");   ← 向手机发送一条数据
31            dos.flush();
32            System.out.println("服务器休眠20 秒......");
```

```
33              Thread.sleep(500);
34              String msg="";
35              if((msg= dis.readUTF())!=null){
36                  System.out.println(msg);
37              }
38              dos.writeUTF("你发来的数据服务器收到了。^_^");  ← 发送第 2 条数据
39              dos.flush();
40          }
41      }
42      catch (Exception e) {System.out.println("读写错误");}
43      finally{
44              try {
45                  dis.close();
46                  dos.close();
47              } catch (IOException e) {e.printStackTrace();}
48          }
49      }
50  }
51  public static void main(String[] args) throws IOException
52  {
53      new Server();
54  }
55  }
```

（3）修改配置文件 AndroidManifest.xml。

在配置文件 AndroidManifest.xml 中加入允许访问网络的权限语句：

```
<uses-permission android:name="android.permission.INTERNET" />
```

程序由客户端程序和服务器端程序两部分组成。

（1）客户端程序。

① 程序的第 28～35 行为设置线程策略，避免程序出现 NetworkOnMainThreadException 异常，这是 Android 与 Java 不同的地方。

② 在第 46 行创建一个可以连接到 server 的套接字，其端口为 4321。运行程序时，当程序执行到该语句，立即向服务器发起连接。

③ 在第 59 行调用 ReadStr()方法,通过数据输入流读取从服务器发送到线路里的信息。

④ 在第 63 行调用 WriteString(String str)方法，通过数据输出流向由套接字建立的连接线路（向服务器端方向）发送信息。

（2）服务器端程序。

① 在第 17 行创建多线程，可以用于多客户端的连接。

② 在第 22 行创建服务器端套接字,设定其端口号为 4321，该端口号与客户机套接字的端口号必须一致。注意，这里要使用 try-catch 结构处理异常事件。

③ 在第 26 行令服务器端套接字对象使用 accept()方法监听端口，等待接收客户机传来的连接信号。

④ 在第 28、29 行建立套接字的数据输入流 dis 及数据输出流 dos。

⑤ 在第 30 行通过数据输出流向由套接字建立的连接线路（向客户端方向）发送"连接已经建立"的信息。

⑥ 在第 35 行通过数据输入流读取客户机发送在线路里的信息。

⑦ 在第 37 行显示接收到的信息。

将服务器端程序保存为 Server.java，编译程序。首先运行服务器程序，然后启动模拟器运行客户端程序。

程序运行结果如图 6.10 所示（先运行服务器端程序，再运行客户端程序）。

(a) 服务器端运行结果

(b) 客户端运行结果

图 6.10　远程数据传输

6.2.3　应用 Callable 接口实现多线程 Socket 编程

1. Callable 接口

在 Java 语言中，经常使用的 Thread 类在 run()方法执行完之后是没有返回值的，要实现子线程完成任务后返回值给主线程需要借助第三方转存。Callable 接口提供了一种有返回值的多线程实现方法。

Callable 接口的定义如下：

```
public interface Callable<V>
{
    V call() throws Exception;
}
```

2. 线程接口 Runnable 和 Callable 的区别

Callable 接口与 Java 的 Runnable 接口的作用很类似，但它们之间又有很多不同，其区别如下。

（1）Runnable 接口自从 Java1.1 就有了，而 Callable 接口是在 Java 1.5 之后新增加的。

（2）Callable 接口中定义的方法是 call()，Runnable 中定义的方法是 run()。

（3）Callable 接口的任务执行后可以有返回值，而 Runnable 接口的任务是不能有返回值的（其返回类型为 void）。

（4）call()方法可以抛出异常，run()方法不能抛出异常。

（5）运行 Callable 接口任务可以返回一个 Future 对象，该对象为异步计算的结果。它提供了检查计算是否完成的方法，以等待计算的完成，并检索计算的结果。用户通过 Future 对象可以了解任务执行情况，可以取消任务的执行，还可以获取执行结果。

3. 应用 Callable 接口实现多线程 Socket 编程示例

【例 6-6】 应用 Callable 接口实现多线程远程数据通信示例。

Android 客户端程序由实现 Callable 接口的 connSocket.java 和主程序 MainActivity.java 两部分组成。

（1）实现 Callable 接口的 connSocket.java。

```
1   package com.example.ex6_6;
2   import java.io.DataInputStream;
3   import java.io.DataOutputStream;
4   import java.net.InetSocketAddress;
5   import java.net.Socket;
6   import java.net.SocketAddress;
7   import java.util.concurrent.Callable;

8   class connSocket implements Callable<String>
9   {
10  static DataInputStream datain;
11  static DataOutputStream dataout;
12  static Socket ss;
13  String IP="58.199.89.161";
14  private String runlog=" ";
15  public String call() throws Exception      ← 定义 Callable 接口的 call()方法
16  {
```

```
17    try {
18      ss = new Socket();
19      SocketAddress socketAddress = new InetSocketAddress(IP, 8888);
20      ss.connect(socketAddress, 5000);  //设置超时时间
21      datain=new DataInputStream(ss.getInputStream());//创建数据输入流
22      dataout=new DataOutputStream(ss.getOutputStream());//创建数据输出流
23      dataout.writeUTF("客户端发来的信息：Socket 我来了！！");
24      this.runlog= datain.readUTF();
25      Thread.sleep(500);
26      dataout.writeUTF("客户端发来的信息：我已经收到服务器的信息！！");
27      this.runlog= datain.readUTF();
28    }catch (Exception e){ this.runlog="Socket 错误";}
29    return  this.runlog;
30  }

31  public  static  void disConnet()
32  {
33    if(datain != null)
34      try{ datain.close(); }catch (Exception e){ e.printStackTrace(); }
35    if(dataout != null)
36      try{ dataout.close();}catch (Exception e){ e.printStackTrace(); }
37    if(ss != null)
38      try{ ss.close();}catch (Exception e){ e.printStackTrace(); }
39  }
40 }
```

（2）主程序 MainActivity.java。

```
1  package com.example.ex6_6;
2  import androidx.appcompat.app.AppCompatActivity;
3  import android.os.Bundle;
4  import android.view.View;
5  import android.widget.Button;
6  import android.widget.ImageView;
7  import android.widget.TextView;
8  import java.util.concurrent.FutureTask;

9  public class MainActivity extends AppCompatActivity
10 {
11   ImageView img;
12   Button connBtn;
13   TextView txt;
14   @Override
15   protected void onCreate(Bundle savedInstanceState)
16   {
17     super.onCreate(savedInstanceState);
18     setContentView(R.layout.activity_main);
```

```
19    img=(ImageView)findViewById(R.id.imageView);
20    img.setImageResource(R.drawable.img2);
21    txt=(TextView)findViewById(R.id.textView);
22    connBtn = (Button)findViewById(R.id.button);
23    connBtn.setOnClickListener(new mClick());
24  }
25  class mClick implements View.OnClickListener
26  {
27   String str;
28   @Override
29   public void onClick(View v)
30   {
31    if(v == connBtn) {
32      connSocket conn = new connSocket();
33      try {                                        ← 需要利用 Future 对象获得执行结果
34         FutureTask<String> msg = new FutureTask<String>(conn);
35         new Thread(msg).start();
36         str = msg.get();           ← 使用 get()方法获取线程的返回值
37         txt.append(str);
38      } catch (Exception e) { txt.setText("连接错误!!!!!! ");}
39      finally { conn.disConnet(); }
40    } // if_end          //使用 get()方法获取线程的返回值
41   } // onClick_end
42  } // mClick_end
43 }
```

（3）修改配置文件 AndroidManifest.xml。

在配置文件 AndroidManifest.xml 中加入允许访问网络的权限语句：

```
<uses-permission android:name="android.permission.INTERNET"/>
```

（4）服务器端程序 Server.java。

```
1   import java.io.DataInputStream;
2   import java.io.DataOutputStream;
3   import java.io.IOException;
4   import java.net.ServerSocket;
5   import java.net.Socket;

6   public class Server {
7    private ServerSocket ss;
8    private Socket socket;
9    private DataInputStream in;
10   private DataOutputStream out;
11   public Server(){
12      new ServerThread().start();
13   }
14   class ServerThread extends Thread{
```

```
15      public void run() {
16          try {
17              ss=new ServerSocket(8888);
18              System.out.println("服务器启动了");
19              while(true){
20                  socket =ss.accept();
21                  System.out.println("有客户端连接到服务器");
22                  in=new DataInputStream(socket.getInputStream());
23                  out=new DataOutputStream(socket.getOutputStream());
24                  String msg="";
25                  if((msg=in.readUTF())!=null){
26                      System.out.println(msg);
27                  }
28                  out.writeUTF("恭喜你，连接服务器成功！    \n");
29                  sleep(500);
30                  if((msg=in.readUTF())!=null){
31                      System.out.println(msg);
32                  }
33                  out.writeUTF("你发来的数据服务器收到了。^_^");
34                  out.flush();
35              }
36          } catch (IOException | InterruptedException e) {
37          e.printStackTrace();
38          }finally{
39          try {
40              in.close();
41              out.close();
42          } catch (IOException e) {
43              e.printStackTrace();
44          }
45          }
46          }
47      }

48      public static void main(String[] args) throws IOException {
49          new Server();
50      }
51  }
```

6.3 基于 HTTP 的网络程序设计

6.3.1 建立 PHP 服务器网站

如今有不少人通过手机上网，"哪里有人群，哪里就有发展"，这也导致互联网正在向

第6章 网络通信

移动端发展。人们使用手机通过无线网络获取 Web 服务器的数据，如图 6.11 所示。

图 6.11 手机获取网络服务器的数据

6.3.2 应用 HttpURLConnection 访问 Web 服务器

1. HttpURLConnection 类

HttpURLConnection 是一种多用途、轻量级的 HTTP 客户端，大多数的应用程序可以使用它来进行 HTTP 操作。但 HttpURLConnection 是 Java 的标准类，没有对其进行封装，需要进行比较复杂的设置，用起来不太方便。

2. StrictMode 类

在 MainActivity 中调用 HttpURLConnection 类的网络操作方法可能会导致 Activity 的一些问题，在 Android 2.3 版本以后系统增加了 StrictMode 类，这个类对网络的访问方式进行了一定的改变。StrictMode 通常用于捕获磁盘访问或者网络访问中与主进程之间交互产生的问题，因为在主进程中 UI 操作和一些动作的执行是最经常用到的，它们之间会产生一定的冲突问题。将磁盘访问和网络访问从主线程中剥离可以使磁盘或者网络的访问更加流畅，从而提升响应度和用户体验。

在使用 HttpURLConnection 前需要调用 HttpURLConnection 的以下两个方法。

（1）StrictMode.setThreadPolicy()：线程对象管理策略。

（2）StrictMode.setVmPolicy()：VM 虚拟机对象管理策略。

StrictMode 方法需要在主页面的 onCreate 方法里添加以下代码。

```
StrictMode.setThreadPolicy(
    newStrictMode.ThreadPolicy.Builder()      //构造 StrictMode 线程对象
        .detectDiskReads()                     //当发生磁盘读操作时输出
        .detectDiskWrites()                    //当发生磁盘写操作时输出
        .detectNetwork()                       //访问网络时输出，包括磁盘读/写和网络 I/O
        .penaltyLog()                          //以日志的方式输出
        .build()
);
```

setVmPolicy 是关于 VM 虚拟机等方面的策略。

```
StrictMode.setVmPolicy(
    new StrictMode.VmPolicy.Builder()         //构造 StrictMode VM 虚拟机对象
        .detectLeakedSqlLiteObjects()         //探测 SQLite 数据库操作
```

```
            .detectLeakedClosableObjects()      //探测关闭操作
            .penaltyLog()                       //以日志的方式输出
            .penaltyDeath()
            .build()
);
```

【例 6-7】 从 Web 服务器读取图像文件。

（1）在 Web 服务器端，将事先准备好的图片 dukou.jpg 保存到 Web 服务器的根目录下。

（2）设计界面布局文件 activity_main.xml。

界面布局设计如图 6.12 所示，设置一个按钮、两个用于显示信息的文本框和一个显示图像的 ImageView。

图 6.12　从 Web 服务器读取图像文件

（3）设计主程序 MainActivity.java。

```
1    package com.example.ex6_7;
2    import androidx.appcompat.app.AppCompatActivity;
3    import android.os.Bundle;
4    import android.graphics.Bitmap;
5    import android.graphics.BitmapFactory;
6    import android.os.Handler;
7    import android.os.Message;
8    import android.os.StrictMode;
9    import android.view.View;
10   import android.widget.Button;
11   import android.widget.EditText;
12   import android.widget.ImageView;
13   import java.io.InputStream;
```

```
14    import java.net.HttpURLConnection;
15    import java.net.URL;

16    public class MainActivity extends AppCompatActivity
17    {
18      ImageView img;
19      TextView txt1, txt2;
20      Button connBtn;
21      HttpURLConnection conn = null ;
22      InputStream inStrem = null;
23      String str="http://58.199.71.24/dukou.jpg"; //Web服务器IP地址
24      HHandler mHandler=new HHandler();
25      @Override
26      protected void onCreate(Bundle savedInstanceState)
27      {
28        super.onCreate(savedInstanceState);
29        setContentView(R.layout.activity_main);
30        img=(ImageView)findViewById(R.id.imageView);
31        txt1=(TextView)findViewById(R.id.textView);
32        txt2=(TextView)findViewById(R.id.textView2);
33        connBtn=(Button)findViewById(R.id.button);
34        connBtn.setOnClickListener(new mClick());
35      }
36    class mClick implements View.OnClickListener
37    {
38      public void onClick(View v)
39      {
40        StrictMode.setThreadPolicy(
41          new StrictMode
42              .ThreadPolicy
43              .Builder()
44              .detectDiskReads()
45              .detectDiskWrites()
46              .detectNetwork()
47              .penaltyLog()
48              .build()
49        );
50        StrictMode.setVmPolicy(
51          new StrictMode
52              .VmPolicy
53              .Builder()
54              .detectLeakedSqlLiteObjects()
55              .detectLeakedClosableObjects()
56              .penaltyLog()
57              .penaltyDeath()
58              .build()
59        );
```

```
60          getPicture();
61       }
62  }

63   private void getPicture(){
64      try {
65         URL url = new URL(str);              //构建图片的URL地址
66         conn = (HttpURLConnection) url.openConnection();
67         conn.setConnectTimeout(5000);        //设置超时的时间为5000ms
68         conn.setRequestMethod("GET");        //设置获取图片的方式为GET
69         if ( conn.getResponseCode() == 200) { //响应码为200则访问成功
70            //获取连接的输入流,这个输入流就是图片的输入流
71            inStrem = conn.getInputStream();
72            Bitmap bmp= BitmapFactory.decodeStream(inStrem);
73            //由于发送的信息不是字符串,不能使用sendMessage(msg)方法
74            mHandler.obtainMessage(0, bmp).sendToTarget();
75            int result = inStrem.read();
76            while (result != -1){
77                txt1.setText((char)result);
78                result = inStrem.read();
79            }
80            //关闭输入流
81            inStrem.close();
82            txt1.setText("(1)建立输入流成功!");
83         }
84      }catch(Exception e2)  { txt1.setText("(3)IO流失败");}
85   } //getPicture()结束

86   /**
87      Android利用Handler来实现UI线程的更新。
88      Handler是Android中的消息发送器,主要接收子线程发送的数据,并用此数据
89      配合主线程更新UI。接收消息,处理消息,此Handler会与当前主线程一起运行
90   **/
91   class HHandler extends Handler
92   {
93    public void handleMessage(Message msg){//子类必须重写此方法,接收数据
94        super. handleMessage( msg);
95        txt2.setText("(2)下载图像成功!");
96        img.setImageBitmap((Bitmap) msg.obj);   //更新UI
97    }
98   }
99  } //主类结束
```

（4）修改配置文件。

① 支持HTTP访问网络。

由于Android 9以后默认仅支持HTTPS访问网络,不支持HTTP,为了也能支持HTTP访问网络,则需要在配置文件中添加以下语句。

```
android:usesCleartextTraffic="true"
```

② 连接共享类库。

连接共享类库的语句如下：

```
<uses-library
    android:name="org.apache.http.legacy"
    android:required="false"
/>
```

③ 添加访问网络权限和访问 HTTP 网站权限：

```
<uses-permission android:name="android.permission.INTERNET" />
```

完整的配置文件 AndroidManifest.xml 代码如下：

```
1   <?xml version="1.0" encoding="utf-8"?>
2   <manifest xmlns:android="http://schemas.android.com/apk/res/android"
3       package="com.example.ex6_7">
4       <application
5           android:allowBackup="true"
6           android:icon="@mipmap/ic_launcher"
7           android:label="@string/app_name"
8           android:roundIcon="@mipmap/ic_launcher_round"
9           android:supportsRtl="true"
10          android:theme="@style/Theme.Ex6_7"
11          android:usesCleartextTraffic="true"
12          >
13          <uses-library
14              android:name="org.apache.http.legacy"
15              android:required="false" />
16          <activity android:name=".MainActivity">
17              <intent-filter>
18                  <action android:name="android.intent.action.MAIN" />
19                  <category android:name="android.intent.category.LAUNCHER"/>
20              </intent-filter>
21          </activity>
22      </application>
23      <uses-permission android:name="android.permission.INTERNET" />
24  </manifest>
```

【例 6-8】 以 GET 方式和 POST 方式对 Web 服务器读取及发送数据。

在手机客户端编写界面布局文件和主程序文件，在 Web 服务器端编写接收 GET 请求的 play-get.php 文件和接收 POST 请求的 play-post.php 文件。

（1）设计界面布局文件 activity_main.xml。界面布局如图 6.13 所示。

（2）设计主程序 MainActivity.java。

图6.13 以 GET 方式和 POST 方式向 Web 服务器发送请求

```
1   package com.example.ex6_8;
2   import androidx.appcompat.app.AppCompatActivity;
3   import android.os.Bundle;
4   import android.os.StrictMode;
5   import android.view.View;
6   import android.widget.Button;
7   import android.widget.EditText;
8   import android.widget.TextView;
9   import java.io.BufferedReader;
10  import java.io.InputStreamReader;
11  import java.io.PrintWriter;
12  import java.net.HttpURLConnection;
13  import java.net.URL;
14  import java.util.ArrayList;
15  import java.util.HashMap;
16  import java.util.List;
17  import java.util.Map;

18  public class MainActivity extends AppCompatActivity {
19      Button getBtn, postBtn;
20      TextView txt;
21      EditText editsid, editname, editemail;
22      @Override
23      protected void onCreate(Bundle savedInstanceState) {
24          super.onCreate(savedInstanceState);
25          setContentView(R.layout.activity_main);
26          getBtn=(Button)findViewById(R.id.button_get);
```

```
27      postBtn=(Button)findViewById(R.id.button_post);
28      editsid=(EditText)findViewById(R.id.editText_ID);
29      editname=(EditText)findViewById(R.id.editText_Name);
30      editemail=(EditText)findViewById(R.id.editText_email);
31      txt=(TextView)findViewById(R.id.textView_txt);
32      setVersion();// 设置线程策略
33      getBtn.setOnClickListener(new mClick());
34      postBtn.setOnClickListener(new mClick());
35   }

36   class mClick implements View.OnClickListener{
37      StringBuilder stringBuilder = new StringBuilder();
38      BufferedReader buffer = null;
39      HttpURLConnection connGET = null;
40      HttpURLConnection connPOST = null;
41      @Override
42      public void onClick(View v) {
43          if(v == getBtn) {
44              //获取界面文本框中的文字内容
45              String sid=editsid.getText().toString();
46              String name=editname.getText().toString();
47              String email=editemail.getText().toString();
48              try{
49                  String str="http://58.199.71.24/test/play-get.php?sid="+
50                          sid+"&name="+name+"&email="+email;
51                  URL url = new URL(str);          //构建 Web 服务器的 URL 地址
52                  connGET = (HttpURLConnection) url.openConnection();
53                  connGET.setConnectTimeout(5000);   //设置超时的时间
54                  connGET.setRequestMethod("GET"); //设置获取数据的方式为 GET
55                  if ( connGET.getResponseCode() == 200) {
56                      buffer = new BufferedReader(new
57                       InputStreamReader(connGET.getInputStream()));
58                      for(String s = buffer.readLine();
59                          s != null;
60                          s = buffer.readLine()){
61                          stringBuilder.append(s);    //构造字符串
62                      }
63                      txt.setText(stringBuilder);
64                      buffer.close();
65                  }
66              }
67              catch(Exception e){
68                  txt.setText("get 提交 err.....");
69              }
70          } //if(v == getBtn)结束
71          if(v == postBtn) {
72              //获取界面文本框中的文字内容
```

```
73          String sid=editsid.getText().toString();
74          String name=editname.getText().toString();
75          String email=editemail.getText().toString();
76          try{
77              String str="http://58.199.71.24/test/play-post.php";
78              URL url = new URL(str);              //构建 Web 服务器的 URL 地址
79              connPOST = (HttpURLConnection) url.openConnection();
80              connPOST.setConnectTimeout(5000);    //设置超时的时间为 5000ms
81              connPOST.setRequestMethod("POST");//设置获取数据的方式为 POST
82              // 发送 POST 请求必须设置以下两行
83              connPOST.setDoOutput(true);
84              connPOST.setDoInput(true);
85              //----------发送数据--------
86              // 建立对应的输出流
87              PrintWriter printWriter =
88                      new PrintWriter(connPOST.getOutputStream());
89              Map<String, Object> paramsMap = new HashMap<String, Object>();
90              paramsMap.put("sid", sid);
91              paramsMap.put("name", name);
92              paramsMap.put("email", email);
93              printWriter.write(paramsMap.toString());  // 发送请求参数
94              printWriter.flush();                      // flush 输出流的缓冲
95              //----------接收数据--------
96              // 定义 BufferedReader 输入流来读取 URL 的返回数据
97              buffer = new BufferedReader(new
98              InputStreamReader(connPOST.getInputStream()));
99              for(String s = buffer.readLine();
100                 s != null;
101                 s = buffer.readLine()){
102                 stringBuilder.append(s);         //构造字符串
103             }
104             txt.setText(stringBuilder);
105             buffer.close();
106         } //try 结束
107         catch(Exception e){ txt.setText("response err....."); }
108     }
109 }
110 }
111 //设置线程策略
112 void setVersion()
113 {
114   StrictMode.setThreadPolicy(new StrictMode
115       .ThreadPolicy.Builder()
116       .detectDiskReads()
117       .detectDiskWrites()
118       .detectNetwork()  //这里若替换为 detectAll()就包括了磁盘读/写和网络 I/O
119       .penaltyLog() //打印 logcat,也可以定位到 dropbox,通过文件保存 log
```

```
120         .build());
121     StrictMode.setVmPolicy(new StrictMode
122         .VmPolicy.Builder()
123         .detectLeakedSqlLiteObjects()  //探测SQLite数据库操作
124         .penaltyLog()  //打印logcat
125         .penaltyDeath()
126         .build());
127     }
128 }
```

（3）修改配置文件，设置网络访问权限和支持HTTP访问权限。

```
1  <?xml version="1.0" encoding="utf-8"?>
2  <manifest xmlns:android="http://schemas.android.com/apk/res/android"
3      package="com.example.ex6_8">
4      <application
5          android:allowBackup="true"
6          android:icon="@mipmap/ic_launcher"
7          android:label="@string/app_name"
8          android:roundIcon="@mipmap/ic_launcher_round"
9          android:supportsRtl="true"
10         android:theme="@style/Theme.ex6_8"
11          android:usesCleartextTraffic="true"
12         >
13         <uses-library
14              android:name="org.apache.http.legacy"
15              android:required="false" />
16         <activity android:name=".MainActivity">
17             <intent-filter>
18                 <action android:name="android.intent.action.MAIN" />
19                 <category android:name="android.intent.category.
                    LAUNCHER" />
20             </intent-filter>
21         </activity>
22     </application>
23      <uses-permission android:name="android.permission.INTERNET" />
24 </manifest>
```

（4）Web服务器端接收GET请求的play-get.php文件。

```
1  <?php
2      header("Content-Type: text/html;charset=utf-8");//设置页面显示的文字编码
3      echo "GET: ";
4      print_r($_GET);
5      $ssid=$_GET["sid"];
6      $sname=$_GET["name"];
7      $semail=$_GET["email"];
8      print_r("GET给服务器的值为数组\n" );
```

```
9      print_r($ssid );
10     print_r("\n");
11     print_r($sname);
12     print_r("\n");
13     print_r($semail);
14     print_r("\n");
15  ?>
```

(5) Web 服务器端接收 POST 请求的 play-post.php 文件。

```
1   <?php
2     header("Content-Type: text/html;charset=utf-8");//设置页面显示的文字编码
3     echo "POST: ";
4     print_r($_POST);
5     $ssid=$_POST["sid"];
6     $sname=$_POST["name"];
7     $semail=$_POST["email"];
8     print_r("POST 给服务器的值为数组\n" );
9     print_r($ssid );
10    print_r("\n");
11    print_r($sname);
12    print_r("\n");
13    print_r($semail);
14    print_r("\n");
15  ?>
```

习题 6

1. 编写一个用户注册程序，向远程服务器注册。
2. 编写一个具有密码验证功能的远程用户登录程序。

第7章 应用Volley框架访问Web服务器

7.1 Volley 框架及其应用

视频讲解

7.1.1 Volley 包的下载与安装

1. Volley 简介

开发 Android 应用项目经常需要用到来自网络的数据,通常应用程序都会使用 HTTP 协议来发送和接收网络数据。Android 系统中主要用 HttpURLConnection 对象来进行 HTTP 通信,几乎在任何项目的代码中都能看到这个类的身影,使用率非常高(早期版本还使用 HttpClient 类,现在已经废弃)。

不过 HttpURLConnection 的用法稍微有些复杂,如果不进行适当封装,很容易就会写出不少重复代码。于是,一些 Android 网络通信框架应运而生。

Android 开发团队意识到有必要将 HTTP 的通信操作再进行简单化,于是在 2013 年推出了一个新的网络通信框架——Volley。Volley 既可以非常简单地进行 HTTP 通信,也可以轻松加载网络上的图片。除了简单、易用之外,Volley 在性能方面也进行了大幅度的调整,它的设计目标是非常适合进行数据量不大但通信频繁的网络操作。

2. 下载和安装 Volley

用户可以到网站 https://www.jb51.net/softs/541659.html 下载 volley.jar。

新建一个 Android 项目,将 Android Studio 系统设置为 Project 模式,然后将 volley.jar 文件复制到 libs 目录下,如图 7.1 所示。

右击新粘贴的 volley.jar 项,在弹出的菜单中选择 Add As Library 命令完成 jar 包的安装,如图 7.2 所示。

7.1.2 JSON 数据格式简介

Volley 在处理数据的接收与发送时使用的是 JSON 数据格式,下面对 JSON 数据格式进行简要介绍。

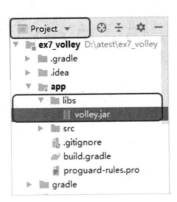
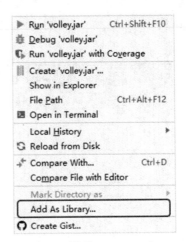

图 7.1　在 Project 模式下复制 Volley 包　　　　图 7.2　安装 volley.jar 包

JSON（JavaScript Object Notation）是一种轻量级的数据交换格式。JSON 采用完全独立于语言的纯文本格式，易于人们阅读和编写，同时也易于机器解析和生成（一般用于提升网络传输速率），从而使 JSON 成为网络传输中理想的数据交换语言。

1. JSON 数据格式

JSON 用"键-值"对形式表示数据，其数据的书写格式如下：

键名(key) : 值(value)

"键-值"对的键名 key 必须是字符串，后面写一个冒号"："，然后是值 value，值 value 可以是字符串、数值、布尔值。

例如：

"firstName" : "John"

这很容易理解，等价于下列 JavaScript 语句：

firstName = "John"

2. JSON 对象

JSON 对象可以包含多个"键-值"对，要求在大括号"{ }"中书写，"键-值"对之间用逗号","分隔。

例如：

{ "firstName":"John" , "lastName":"Doe" , "age":20 }

这一点也很容易理解，等价于下列 JavaScript 语句：

firstName = "John"
lastName = "Doe"
age = 20

JSON 对象的值也可以是另一个对象，例如：

```
{
    "Name":"John" ,
    "age": 20 ,
    "hobby":"打篮球",
    "friend":{"Name":"Suny" ,"age":19 ,"hobby":"看书"}
}
```

3. JSON 数组

JSON 数组可以包含多个 JSON 数据作为数组元素，每个元素之间用逗号","分隔，要求在中括号"[ ]"中书写。

例如：

```
var meber = [ "John" , 20 ,"打篮球" ];
```

JSON 数组的元素可以包含多个对象，例如：

```
var employees = [
    { "firstName":"John" , "lastName":"Doe" },
    { "firstName":"Anna" , "lastName":"Smith" },
    { "firstName":"Peter" , "lastName":"Jones" }
];
```

可以像这样访问 JavaScript 对象数组中的第一个元素：

```
employees[0].lastName;
```

其返回值为 Doe。

用户也可以修改其数据：

```
employees[0].lastName = "Jobs";
```

4. JSON 文件

JSON 文件的类型是 .json，可以用记事本或其他编辑工具编写 JSON 文件。

5. 解析 JSON 数据

Android 解析 JSON 格式数据需要使用 JSONObject 对象和 JSONArray 对象，下面通过一个示例说明 Android 解析 JSON 格式数据的方法。

【例 7-1】 解析 JSON 格式数据示例。

（1）界面布局如图 7.3 所示。

（2）主控制程序 MainActivity.java 代码如下：

```
1    package com.example.ex7_json;
2    import androidx.appcompat.app.AppCompatActivity;
3    import android.os.Bundle;
```

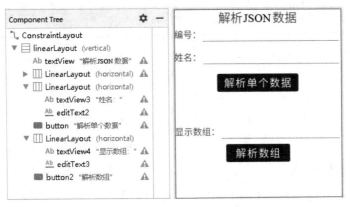

图 7.3　解析 JSON 格式数据的界面布局

```
4    import android.view.View;
5    import android.widget.Button;
6    import android.widget.EditText;
7    import org.json.JSONArray;
8    import org.json.JSONException;
9    import org.json.JSONObject;
10   public class MainActivity extends AppCompatActivity
11   {
12       EditText txt1, txt2, txt3;
13       Button   jsonBtn, arrayBtn;
14       /*  设有 JSON 数据
15       JSONArray jdata = [{"sid":1001, "name":"张大山"},
16                          {"sid":1002, "name":"李小丽"} ];
17       */
18       JSONObject jid,jname;
19       @Override
20       protected void onCreate(Bundle savedInstanceState) {
21           super.onCreate(savedInstanceState);
22           setContentView(R.layout.activity_main);
23           txt1 = (EditText)findViewById(R.id.editText);
24           txt2 = (EditText)findViewById(R.id.editText2);
25           txt3 = (EditText)findViewById(R.id.editText3);
26           jsonBtn = (Button)findViewById(R.id.button);
27           arrayBtn = (Button)findViewById(R.id.button2);
28           jsonBtn.setOnClickListener(new mClick());
29           arrayBtn.setOnClickListener(new mClick());
30       }
31       void setJsonData()  {
32         try {
33             jid = new JSONObject();
```

```
34          jname = new JSONObject();
35          jid.put("sid",1001);
36          jname.put("name","张大山");
37          String sid=jid.getString("sid");
38          txt1.setText(sid);
39          String sname = jname.getString("name");
40          txt2.setText(sname);
41      }catch (JSONException e){  }
42   }

43   void setArrayData(){
44     try {
45          JSONArray jdata = new JSONArray();
46          JSONObject p1 = new JSONObject();
47          JSONObject p2 = new JSONObject();
48          p1.put("sid",1001).put("name","张大山");
49          p2.put("sid",1002).put("name","李小丽");
50          jdata.put(p1);
51          jdata.put(p2);
52          String sid , sname;
53          int length = jdata.length();
54          for(int i=0; i<length; i++){   //遍历JSONArray
55              JSONObject jsonObject = jdata.getJSONObject(i);
56              sid = jsonObject.getString("sid") + ":";
57              sname = jsonObject.getString("name") + "\n";
58              txt3.append(sid + sname);
59          }
60      }catch (JSONException e){  }
61   }

62   public class mClick implements View.OnClickListener
63   {
64      @Override
65      public void onClick(View v) {
66          if(v == jsonBtn)  setJsonData();
67          else if(v == arrayBtn) setArrayData();
68      }
69    }
70 }
```

程序运行结果如图 7.4 所示。

图 7.4 解析 JSON 数据程序的运行结果

7.1.3 Volley 的工作原理和几个重要对象

1. Volley 的基本工作原理

Volley 在工作时,首先由主线程(应用程序)发起一条 HTTP 请求,将请求添加到缓存队列中,然后通过缓存调度线程从缓存队列中取出一个请求,在缓存中解析并做出响应,最后将解析后的响应发送给主线程。在 Volley 内部创建两个线程,一个为缓存调度线程,另一个为网络调度线程,优先在缓存中解析并响应请求,如果缓存不能解析,则由网络线程使用 HTTP 发送请求给远程 Web 服务器解析。Volley 的基本工作原理如图 7.5 所示。

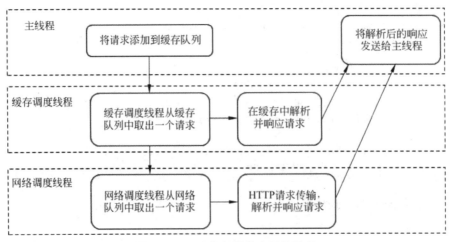

图 7.5 Volley 框架的基本工作原理

2. Volley 的几个重要对象

使用 Volley 框架需要创建以下几个重要对象。
（1）RequestQueue：用来执行请求的请求队列。
（2）Request：用来构造一个请求对象。
Request 对象主要有以下几种类型。
- StringRequest：响应的主体为字符串。
- JsonArrayRequest：发送和接收 JSON 数组。
- JsonObjectRequest：发送和接收 JSON 对象。
- ImageRequest：发送和接收 Image 对象。

7.1.4　Volley 的基本使用方法

Volley 的用法非常简单，设应用程序发起一条 HTTP 请求，然后接收 HTTP 响应，其步骤如下。
（1）创建一个 RequestQueue 对象，可以调用以下方法得到。

```
RequestQueue mQueue = Volley.newRequestQueue(context);
```

注意这里得到的 RequestQueue 是一个请求队列对象，它可以缓存所有的 HTTP 请求，然后按照一定的算法并发地发出这些请求。

RequestQueue 内部的设计是非常适合高并发的，因此不必为每一次 HTTP 请求都创建一个 RequestQueue 对象，否则非常浪费资源，基本上在每一个需要和网络交互的 Activity 中创建一个 RequestQueue 对象就可以了。

（2）为了要发出一条 HTTP 请求，还需要创建一个 JsonObjectRequest 对象，例如：

```
JsonObjectRequest postRequest = new JsonObjectRequest(
Request.Method.POST,                 //第1个参数，提交请求方式，这里为POST方式
"http://192.168.1.5/postData.php" ", //第2个参数，请求的服务器地址
new JSONObject(params),              //第3个参数，向服务器发送数据
new Response.Listener<JSONObject >() //第4个参数，响应正确时的处理
{
   @Override
   public void onResponse(JSONObject response){
       //onResponse 获取服务器的响应，response 为返回的响应值
    }
},
new Response.ErrorListener()     //第5个参数，响应错误时的反馈信息
{
    @Override
    public void onErrorResponse(VolleyError error)
    { Log.e("TAG", error.getMessage(), error);   }
});
```

可以看到，这里出现了一个 JsonObjectRequest 对象，JsonObjectRequest 的构造函数需

要传入 5 个参数：

第 1 个参数为请求方式；

第 2 个参数为目标服务器的 URL 地址，注意，这个地址不能使用 127.0.0.1 或 localhost，而需要使用本机的 IP 地址；

第 3 个参数为发送给服务器的数据；

第 4 个参数为服务器响应成功的回调；

第 5 个参数为服务器响应失败的回调。

（3）将这个 JsonObjectRequest 对象添加到 RequestQueue，例如：

```
mQueue.add(stringRequest);
```

需要注意的是，使用 Volley 框架一定要在 AndroidManifest.xml 文件中添加网络权限和支持 HTTP 访问权限：

```
<uses-permission android:name="android.permission.INTERNET" />
<intent-filter><action android:name="android.intent.action.MAIN" />
```

并将标签<application>的 android:usesCleartextTraffic 属性设置为 true。

【例 7-2】 假设有 JSON 格式数据{"sid":1001, "name":"张大山"}，应用 Volley 框架使用 POST 方式向 Web 服务器提交该数据。

该程序分为手机端应用程序和服务器端的 JSON 文件，现设计如下。

（1）在 Web 服务器端建立 postData.php 文件。在 Web 服务器的 www 根目录下新建 test 目录，并在 test 目录中建立 postData.php 文件，其文件代码如下：

```
1    <?php
2    header('Content-Type: text/html; charset=utf-8');
3    $json = file_get_contents("php://input");   //接收数据
4    $jsondata = json_decode($json, true);  //对 JSON 格式解码，并转换为 PHP 变量
5    $sid = $jsondata['sid'];     //取出提交的数据值，可以交给后续处理
6    $name = $jsondata['name'];
7    echo '{"sid":"'.$sid.'","name":"'.$name.'"}';//构造出 JSON 格式，返回给客户端
8    ?>
```

（2）把 volley.jar 复制到项目的 app\libs 目录下，并完成 jar 包的安装。

（3）设计手机端界面布局程序。手机端界面布局程序如图 7.6 所示。

（4）设计主程序 MainActivity.java。

```
1    package com.example.ex7_volley;
2    import androidx.appcompat.app.AppCompatActivity;
3    import android.os.Bundle;
4    import android.view.View;
5    import android.widget.Button;
6    import android.widget.TextView;

7    import com.android.volley.Request;
8    import com.android.volley.RequestQueue;
```

第7章 应用Volley框架访问Web服务器

图 7.6 手机端界面布局设计

```
9   import com.android.volley.Response;
10  import com.android.volley.VolleyError;
11  import com.android.volley.toolbox.JsonObjectRequest;
12  import com.android.volley.toolbox.StringRequest;
13  import com.android.volley.toolbox.Volley;

14  import org.json.JSONException;
15  import org.json.JSONObject;
16  import java.util.HashMap;
17  import java.util.Map;

18  public class MainActivity extends AppCompatActivity {
19  TextView txt;
20  Button btn;
21  Datainfo datainfo = new Datainfo();
22  @Override
23  protected void onCreate(Bundle savedInstanceState) {
24      super.onCreate(savedInstanceState);
25      setContentView(R.layout.activity_main);
26  txt = (TextView)findViewById(R.id.textView);
27  btn = (Button)findViewById(R.id.button);
28  btn.setOnClickListener(new mClick());
29  }

30  public class Datainfo
31  {
32    private String name;
```

```java
33    private String sid;
34    public void setName(String name){this.name = name;}
35    public void setSid(String sid){this.sid = sid;}
36    public String getName(){return name;}
37    public String getSid(){return sid;}
38  }
39  class mClick implements View.OnClickListener{
40    @Override
41    public void onClick(View v) {
42      postServerData();
43      txt.setText(datainfo.getSid()+" "+datainfo.getName());
44    }
45  }
46  void postServerData(){
47    Map<String, String> params = new HashMap<String, String>();//创建Map对象
48    params.put("sid", "1001");
49    params.put("name", "张大山");
50    String jsonURL = "http://10.97.56.251/test/postData.php";
51    RequestQueue mQueue = Volley.newRequestQueue(MainActivity.this);
52    JsonObjectRequest postRequest = new JsonObjectRequest(
53        Request.Method.POST,              //第1个参数，POST方法
54        jsonURL,                          //第2个参数，请求的网址
55        new JSONObject(params),           //第3个参数，向服务器发送数据
56        new Response.Listener<JSONObject>(){  //第4个参数，响应正确时的处理
57        @Override
58        public void onResponse(JSONObject response){
59          try{
60
61            txt.setText(response.toString());  //查看返回的数据
62            String sid = new String(response.getString("sid"));
63            String sname=new String(response.getString("name"));
64            datainfo.setSid(sid);
65            datainfo.setName(sname);
66          }catch (JSONException e) { txt.setText("提交错误"); }
67        } //onResponse()_end
68      }, //Response.Listener<JSONObject>()_end
69  new Response.ErrorListener() {     //第5个参数，响应错误时的反馈信息
70      @Override
71      public void onErrorResponse(VolleyError error) {
72          txt.setText("错误处理");
73      }
74    }
75    ){ }; //JsonObjectRequest()_end
76    mQueue.add(postRequest);
77  } //postServerData()_end
78  }
```

（5）修改配置文件，设置网络访问权限和支持 HTTP 访问权限。

完整的配置文件 AndroidManifest.xml 代码如下：

```xml
1   <?xml version="1.0" encoding="utf-8"?>
2   <manifest xmlns:android="http://schemas.android.com/apk/res/android"
3       package="com.example.ex7_volley">
4     <application
5         android:allowBackup="true"
6         android:icon="@mipmap/ic_launcher"
7         android:label="@string/app_name"
8         android:roundIcon="@mipmap/ic_launcher_round"
9         android:supportsRtl="true"
10        android:theme="@style/Theme.ex7_volley"
11        android:usesCleartextTraffic="true"
12        >
13        <uses-library
14            android:name="org.apache.http.legacy"
15            android:required="false" />
16        <activity android:name=".MainActivity">
17            <intent-filter>
18                <action android:name="android.intent.action.MAIN" />
19                <category android:name="android.intent.category.LAUNCHER" />
20            </intent-filter>
21        </activity>
22    </application>
23    <uses-permission android:name="android.permission.INTERNET" />
24  </manifest>
```

运行程序时，首先启动 Web 服务器，然后运行手机端应用程序。程序运行结果如图 7.7 所示。

图 7.7　应用 Volley 远程向 Web 服务器提交 JSON 数据

7.2 应用 Volley 框架设计网络音乐播放器

下面介绍如何应用 Volley 框架设计网络音乐播放器。

【例 7-3】 音乐信息保存在远程服务器的一个 JSON 格式文件中，读取 JSON 数据内容，把音乐的相关信息显示到屏幕上，并播放音乐。

（1）将 volley.jar 包复制到项目的 libs 目录下，右击新粘贴的 volley.jar 项，在弹出的菜单中选择 Add As Library 选项，安装 volley.jar 包。

（2）将一个音乐文件 mtest.mp3 复制到 Web 服务器根目录 www\music 目录下。现用记事本新建一个 JSON 格式数据文件用于存放音乐资源信息，文件保存到远程 Web 服务器 www\music\ music_info.json 中，其数据如下：

```
{"name":"浪子闲话","singer":"花僮","mp3":"mtest.mp3"}
```

注意：music_info.json 需要保存为 utf-8 格式。在记事本中选择"另存为"命令，将编码格式更改为 utf-8 即可。

（3）读取远程服务器文件内容的核心语句。

```
RequestQueue mQueue = Volley.newRequestQueue(MainActivity.this);
JsonObjectRequest jsonObjectRequest = new JsonObjectRequest(
    jsonUrl,    //第 1 个参数，请求的网址
    null,       //第 2 个参数
    new Response.Listener<JSONObject>() {    //第 3 个参数，响应正确时的处理
        @Override
        public void onResponse(JSONObject response) {
        try {
            String sname=new String(response.getString("name"));
            String ssinger=new String(response.getString("singer"));
            String smp3=new String(response.getString("mp3"));
            musicinfo.setName(sname);
            musicinfo.setSinger(ssinger);
            musicinfo.setMp3(smp3);
        }catch (JSONException e){ Log.e("json 错误", e.getMessage(), e);
        }
      }
    },
    new Response.ErrorListener() {           //第 4 个参数，响应错误时的反馈信息
        @Override
        public void onErrorResponse(VolleyError error)
        { Log.e("TAG 错误", error.getMessage(), error); }
    }
);
mQueue.add(jsonObjectRequest);
```

（4）设计界面布局。

在界面布局文件中，设置了一个显示音乐信息的列表组件 listView，另外 3 个按钮分别用于读取远程服务器上的音乐信息文件、播放音乐及停止播放，并设置 linearLayout(horizontal) 的 gravity 属性值为 center，使这 3 个按钮居中，如图 7.8 所示。

图 7.8　远程服务器音乐播放器界面布局

（5）修改配置文件，设置网络访问权限。

```
1   <?xml version="1.0" encoding="utf-8"?>
2   <manifest xmlns:android="http://schemas.android.com/apk/res/android"
3       package="com.example.ex7_Music">
4       <application
5           android:allowBackup="true"
6           android:icon="@mipmap/ic_launcher"
7           android:label="@string/app_name"
8           android:roundIcon="@mipmap/ic_launcher_round"
9           android:supportsRtl="true"
10          android:theme="@style/Theme.ex7_Music">
11          android:usesCleartextTraffic="true">
12          <uses-library android:name="org.apache.http.legacy"
13              android:required="false" />
14          <activity android:name=".MainActivity">
15              <intent-filter>
16                  <action android:name="android.intent.action.MAIN" />
17                  <category android:name="android.intent.category.LAUNCHER" />
```

```
18            </intent-filter>
19        </activity>
20    </application>
21    <uses-permission android:name="android.permission.INTERNET" />
22 </manifest>
```

（6）编写获取音乐信息类文件 Musicinfo.java。

```
1  package com.example.ex7_Music;
2  public class Musicinfo
3  {
4    private String name;
5    private String singer;
6    private String mp3;
7    public void setName(String name){this.name = name;}
8    public void setSinger(String singer){this.singer = singer;}
9    public void setMp3(String mp3){this.mp3 = mp3;}
10   public String getName(){return name;}
11   public String getSinger(){return singer;}
12   public String getMp3(){return mp3;}
13 }
```

（7）编写主程序 MainActivity.java。

```
1  package com.example.ex7_Music;
2  import androidx.appcompat.app.AppCompatActivity;
3  import android.media.MediaPlayer;
4  import android.os.Bundle;
5  import android.util.Log;
6  import android.view.View;
7  import android.widget.ArrayAdapter;
8  import android.widget.Button;
9  import android.widget.ListView;
10 import android.widget.TextView;
11 import com.android.volley.RequestQueue;
12 import com.android.volley.Response;
13 import com.android.volley.NetworkResponse;
14 import com.android.volley.ParseError;
15 import com.android.volley.VolleyError;
16 import com.android.volley.toolbox.HttpHeaderParser;
17 import com.android.volley.toolbox.JsonObjectRequest;
18 import com.android.volley.toolbox.Volley;
19 import org.json.JSONException;
20 import org.json.JSONObject;
21 import java.io.IOException;
22 import java.io.UnsupportedEncodingException;

23 public class MainActivity extends    AppCompatActivity {
24   TextView txt;
25   ListView  musicView;                    //定义列表组件，用于显示音乐信息
```

```
26    ArrayAdapter<String> adapter;      //定义适配器
27    Button readBtn, playBtn, stopBtn;
28    String[] list={"","",""};
29    Musicinfo musicinfo = new Musicinfo();  //创建输入/输出数据对象
30    String musicUil="http://58.199.89.161/music/";
31    MediaPlayer mMediaPlayer;
32    @Override
33    protected void onCreate(Bundle savedInstanceState) {
34      super.onCreate(savedInstanceState);
35      setContentView(R.layout.activity_main);
36      txt = (TextView)findViewById(R.id.textView2);
37      musicView = (ListView)findViewById(R.id.listView);
38     readBtn = (Button)findViewById(R.id.button);
39     playBtn = (Button)findViewById(R.id.button2);
40     stopBtn = (Button)findViewById(R.id.button3);
41     getServerData();
42     readBtn.setOnClickListener(new mread());
43     playBtn.setOnClickListener(new mplay());
44     stopBtn.setOnClickListener(new mstop());
45    }
46   class mread implements View.OnClickListener {
47     public void onClick(View v) {
48      list[0] = "音乐名称: " + musicinfo.getName();
49      list[1] = "歌手: " + musicinfo.getSinger();
50      list[2] = "音乐文件名: " + musicinfo.getMp3();
51      adapter = new ArrayAdapter<String>(
52        MainActivity.this,
53        android.R.layout.simple_list_item_1,
54        list);
55      musicView.setAdapter(adapter);
56    }
57  }
58  class mplay implements View.OnClickListener{
59     public void onClick(View v) {
60     try {
61       String path = musicUil + musicinfo.getMp3();
62       txt.setText(path);
63       mMediaPlayer = new MediaPlayer();
64       mMediaPlayer.reset();
65       mMediaPlayer.setDataSource(path);
66       mMediaPlayer.prepare();
67       mMediaPlayer.start();
68     }catch (IOException e){    }
69    }
70   }
71   class mstop implements View.OnClickListener{
72    @Override
73    public void onClick(View v) {
```

```
74      if (mMediaPlayer.isPlaying()){
75        mMediaPlayer.stop();
76        mMediaPlayer.reset();
77        mMediaPlayer.release();
78      }
79    }
80  }

81  public void getServerData(){
82    String jsonUrl = musicUil + "music_info.json";
83    RequestQueue mQueue = Volley.newRequestQueue(MainActivity.this);
84    JsonObjectRequest jsonObjectRequest = new JsonObjectRequest(
85        jsonUrl,//第 1 个参数,请求的网址
86        null, //第 2 个参数
87        new Response.Listener<JSONObject>(){ //第 3 个参数,响应正确时的处理
88          @Override
89          public void onResponse(JSONObject response) {
90            try {
91              String sname=new String(response.getString("name"));
92              String ssinger=new String(response.getString("singer"));
93              String smp3=new String(response.getString("mp3"));
94              musicinfo.setName(sname);
95              musicinfo.setSinger(ssinger);
96              musicinfo.setMp3(smp3);
97            }catch (JSONException e){ txt.setText("list 错误");
98              Log.e("json 错误", e.getMessage(), e);
99            }
100         }
101       },
102       new Response.ErrorListener() {    //第 4 个参数,响应错误时的反馈信息
103         @Override
104         public void onErrorResponse(VolleyError error)
105           { Log.e("TAG 错误", error.getMessage(), error); }
106     } )   // 第 84 行 JsonObjectRequest()的右括号
107       { //将 Volley 默认的 ISO-8859-1 格式转换为 utf-8 格式
108     @Override
109     protected Response<JSONObject> parseNetworkResponse(
110         NetworkResponse response) {
111       try { //jsonObject 要和前面的类型一致,此处都是 JSONObject
112         JSONObject jsonObject = new JSONObject(
113           new String(response.data, "UTF-8"));
114         return Response.success(jsonObject,
115         HttpHeaderParser.parseCacheHeaders(response));
116       } catch (UnsupportedEncodingException e) {
117         return Response.error(new ParseError(e));
118       } catch (Exception je) {
119         return Response.error(new ParseError(je)) ;
120       }
121     }
```

解决汉字乱码

```
122     };
123     mQueue.add(jsonObjectRequest);
124   }   // getServerData(){  }结束

125 }
```

程序运行结果如图 7.9 所示。

图 7.9　远程服务器音乐播放器

7.3　访问远程数据库

Android 访问远程数据库,通常的方法是访问远程服务器,由服务器程序连接后台数据库。

7.3.1　把数据写入远程数据库

下面通过应用示例来说明访问远程数据库的方法。

【例 7-4】　编写一个提交用户数据到远程数据库的应用程序。

1）创建数据库

设有 MySQL 数据库,数据库名为 testdb。该数据库中有用户信息数据表 user,其数据结构如表 7-1 所示。

表 7-1　数据表 user 的数据结构

字　段　名	字　段　类　型	说　　明
sid	int(5)	序号
name	Varchar(10)	姓名
email	Varchar(25)	邮箱

2）设计后台服务器端程序

编写后台服务器端程序，该程序需要实现以下两个功能。
- 接收来自手机端提交的数据。
- 把接收到的数据写入 MySQL 数据库。

服务器端程序 postData.php 的代码如下。

```php
1   <?php
2       header('Content-Type: text/html; charset=utf-8');
3       $json = file_get_contents("php://input");//接收发送来的数据
4       $jsondata = json_decode($json, true);
5       $sid = $jsondata['sid'];      //取出数据，交给后面语句写入数据库
6       $name = $jsondata['name'];
7       $email = $jsondata['email'];
8       echo '{"sid":"'.$sid.'","name":"'.$name.'","email":"'.$email.'"}';
9       $db_name="mydatabase";        //数据库名称
10      $user="root";                 //用户
11      $password="";                 //密码
12      $host="localhost";            //服务器地址
13      $dsn="mysql:host=$host;dbname=$db_name";//连接数据库语句
14      try{
15          $con = new PDO($dsn, $user, $password);   //执行连接
16      }catch(Exception $e){   }
17      $con -> query("set names utf8"); //设置数据库的编码
18      $insert_sql= "insert into user(sid,name,email)
19      values('{$sid}', '{$name}','{$email}')";
20      $con -> exec($insert_sql);       // 执行插入数据操作
21  ?>
```

（接收数据；连接数据库；写入数据库）

3）设计 Android 端程序

（1）建立应用程序项目，并安装 volley.jar 包。

（2）设计界面布局。

在界面中设置两个按钮和两个文本标签，如图 7.10 所示。

图 7.10　提交数据到远程数据库的界面布局

（3）设计主控制程序 MainActivity.java。

```java
1   package com.example.ex7_4;
2   import androidx.appcompat.app.AppCompatActivity;
3   import android.os.Bundle;
4   import android.view.View;
```

```java
5   import android.widget.Button;
6   import android.widget.TextView;

7   import com.android.volley.Request;
8   import com.android.volley.RequestQueue;
9   import com.android.volley.Response;
10  import com.android.volley.VolleyError;
11  import com.android.volley.toolbox.JsonObjectRequest;
12  import com.android.volley.toolbox.StringRequest;
13  import com.android.volley.toolbox.Volley;

14  import org.json.JSONException;
15  import org.json.JSONObject;
16  import java.util.HashMap;
17  import java.util.Map;

18  public class MainActivity extends AppCompatActivity {
19      TextView txt1, txt2;
20      Button connBtn,postBtn;
21      Datainfo datainfo = new Datainfo();
22      @Override
23      protected void onCreate(Bundle savedInstanceState) {
24          super.onCreate(savedInstanceState);
25          setContentView(R.layout.activity_main);
26          txt1 = (TextView)findViewById(R.id.textView);
27          txt2 = (TextView)findViewById(R.id.textView2);
28          connBtn = findViewById(R.id.button1);
29          connBtn.setOnClickListener(new mClick1());
30          postBtn = (Button) findViewById(R.id.button);
31          postBtn.setOnClickListener(new mClick());
32      }
33      public class Datainfo
34      {
35          private String name;
36          private String sid;
37          private String email;
38          public void setName(String name){this.name = name;}
39          public void setSid(String sid){this.sid = sid;}
40          public void setEmail(String email){this.email = email;}
41          public String getName(){return name;}
42          public String getSid(){return sid;}
43          public String getEmail(){return email;}
44      }
45      class mClick1 implements View.OnClickListener {
46          @Override
47          public void onClick(View v) {
48              postServerData();
```

```
49        }
50    }
51    class mClick implements View.OnClickListener{
52        @Override
53        public void onClick(View v) {
54            txt2.setText(datainfo.getSid()+"  "+
55                    datainfo.getName()+"  "+
56                    datainfo.getEmail() );
57        }
58    }
59    void postServerData(){
60        Map<String, String> params = new HashMap<String, String>();
61        params.put("sid", "1003");
62        params.put("name", "赵雨雷");
63        params.put("email", "zys@333.com");
64        String jsonURL = "http://10.97.56.251/test/postData.php";
65        try {
66            RequestQueue mQueue = Volley.newRequestQueue(MainActivity.this);
67            JsonObjectRequest postRequest = new JsonObjectRequest(
68                    Request.Method.POST,      //第 1 个参数，POST 方法
69                    jsonURL,                  //第 2 个参数，请求的网址
70                    new JSONObject(params),   //第 3 个参数，向服务器发送数据
71                    new Response.Listener<JSONObject>() {
                                                //第 4 个参数，响应正确时的处理
72                        @Override
73                        public void onResponse(JSONObject response) {
74                            txt2.setText(response.toString()); //查看返回数据
75                            try{
76                                String sid = response.getString("sid");
77                                String sname=response.getString("name");
78                                String semail= response.getString("email");
79                                txt1.setText("提交数据成功");
80                                datainfo.setSid(sid);
81                                datainfo.setName(sname);
82                                datainfo.setEmail(semail);
83                            } catch (JSONException e) {
84                                System.out.println("返回错误");
85                                txt1.setText("提交返回错误");
86                            }
87                        }
88                    },
89                    new Response.ErrorListener() {
                                                //第 5 个参数，响应错误时的反馈信息
90                        @Override
91                        public void onErrorResponse(VolleyError error) {
92                            txt1.setText("错误处理");
93                        }
```

（4）修改配置文件，设置网络访问权限和支持 HTTP 访问权限。

完整的配置文件 AndroidManifest.xml 代码如下。

```
1   <?xml version="1.0" encoding="utf-8"?>
2   <manifest xmlns:android="http://schemas.android.com/apk/res/android"
3       package="com.example.ex7_4">
4       <application
5           android:allowBackup="true"
6           android:icon="@mipmap/ic_launcher"
7           android:label="@string/app_name"
8           android:roundIcon="@mipmap/ic_launcher_round"
9           android:supportsRtl="true"
10          android:theme="@style/Theme.ex7_4"
11          android:usesCleartextTraffic="true"
12          >
13          <uses-library
14              android:name="org.apache.http.legacy"
15              android:required="false" />
16          <activity android:name=".MainActivity">
17              <intent-filter>
18                  <action android:name="android.intent.action.MAIN" />
19                  <category android:name="android.intent.category.LAUNCHER"/>
20              </intent-filter>
21          </activity>
22      </application>
23      <uses-permission android:name="android.permission.INTERNET" />
24  </manifest>
```

启动 Web 服务器后运行程序，首先单击"连接服务器"按钮提交数据到远程服务器，然后单击"查看返回数据"按钮，可以看到从服务器端返回的信息，如图 7.11 所示。

图 7.11 提交数据到远程服务器

■ 7.3.2 读取远程数据库数据

【例 7-5】 编写一个读取远程数据库数据的应用程序。

视频讲解

（1）设有 MySQL 数据库，数据库名为 testdb。该数据库中有数据表 user,其数据表结构同例 7-4。

（2）服务器端连接数据库的 PHP 文件 conn_testdb.php。

```php
1   <?php
2     header('Content-Type: text/html; charset=utf-8');
3     $db_name="mydatabase";       //数据库名称
4     $user="root";                //用户
5     $password="";                //密码
6     $host="localhost";           //服务器地址
7     $dsn="mysql:host=$host;dbname=$db_name";   //连接数据库语句
8     try{
9         $con = new PDO($dsn, $user, $password);  //执行连接
10    }catch(Exception $e){   }
11    $con -> query("set names utf8");//设置数据库的编码

12    $sql ="select * from user";
13    $result = $con -> query($sql);
14    $str="[";
15    if($con)
16    {
17      $i = 0;          // 用来判断是否为第一条数据
18      foreach($result as $row)
19      {
20        if($i != 0){ $str = $str . ","; }
21        else{  $i = 1;   }
22        $str = $str . '{ "sid":"'.$row['sid'].
23                  '", "name":"'.$row['name'].
24                  '", "email": '.$row['email'].'}';
25      }
26    }else
27    {
28      // 如果连接数据库失败，仍然可以返回一条 JSON 数据
29      echo '{ "user_id":101,"name":"服务器出错了","email":"123@abc"}';
30    }
31    echo $str."]";    // 返回 JSON 格式数组到手机端
32  ?>
```

（3）设计界面布局。

在界面中设置一个按钮 Button 和一个列表组件 ListView。

（4）把 volley.jar 复制到项目的 app\libs 目录下，并完成 jar 包的安装。

（5）设计主程序 MainActivity.java。

```
1   package com.example.ex7_5;
2   importandroidx.appcompat.app.AppCompatActivity;
3   import android.os.Bundle;
4   import android.util.Log;
```

```java
5   import android.view.View;
6   import android.widget.ArrayAdapter;
7   import android.widget.Button;
8   import android.widget.ListView;
9   import android.widget.TextView;
10  import com.android.volley.RequestQueue;
11  import com.android.volley.Response;
12  import com.android.volley.VolleyError;
13  import com.android.volley.toolbox.StringRequest;
14  import com.android.volley.toolbox.Volley;
15  import org.json.JSONArray;
16  import org.json.JSONException;
17  import org.json.JSONObject;
18  public class MainActivity extends AppCompatActivity {
19  TextView txt;
20  Button volleyBtn;
21  ListView listdata;
22  Datainfo datainfo=new Datainfo();
23  @Override
24  protected void onCreate(Bundle savedInstanceState) {
25      super.onCreate(savedInstanceState);
26      setContentView(R.layout.activity_main);
27      volleyBtn=(Button)findViewById(R.id.button);
28      txt = (TextView) findViewById(R.id.textView);
29      listdata = (ListView)findViewById(R.id.listView);
30      getServerData();
31    volleyBtn.setOnClickListener(new mClick());
32  }

33  public class Datainfo     //该类用于数据的输入和输出
34  {
35   private String sid;       //序号
36   private String name;      //姓名
37   private String email;     //邮箱
38   public void setSid(String sid){this.sid = sid;}
39   public void setName(String name){this.name = name;}
40   public void setEmail(String email){this.email = email;}
41   public String getName(){return name;}
42   public String getSid(){return sid;}
43   public String getEmail(){return email;}
44  }

45  class mClick implements View.OnClickListener
46  {
47    String[] list={"","",""};
48     @Override
```

```java
49    public void onClick(View v) {
50      list[0] = "序号: " + datainfo.getSid();
51      list[1] = "姓名: " + datainfo.getName();
52      list[2] = "邮箱: " + datainfo.getEmail();
53      ArrayAdapter<String> adapter = new ArrayAdapter<String>(
54              MainActivity.this,
55              android.R.layout.simple_list_item_1,
56              list);
57      listdata.setAdapter(adapter);
58    }  //onClick()_end
59  }  //class mClick_end

60  public void getServerData(){
61    String jsonURL = "http://58.199.89.161/test/conn_testdb.php";
62    try {
63      RequestQueue mQueue = Volley.newRequestQueue(MainActivity.this);
64      StringRequest stringRequest = new StringRequest(
65        jsonURL,//第1个参数，请求的网址
66        new Response.Listener<String>() { //第2个参数，响应正确时的处理
67          @Override
68          public void onResponse(String response) {
69            if(response != null){
70              try {
71                JSONArray jsonArray = new JSONArray(response);
72                for (int i = 0; i < jsonArray.length(); i++) {
73                  JSONObject  jsonData= jsonArray.getJSONObject(i);
74                  String sid = (String)jsonData.get("sid");
75                  String sname = (String)jsonData.get("name");
76                  String semail = (String)jsonData.get("email");
77                  datainfo.setSid(sid);
78                  datainfo.setName(sname);
79                  datainfo.setEmail(semail);
80                }
81              }catch (JSONException e) {
82                txt.setText("list 错误");
83                Log.e("json 错误", e.getMessage(), e);
84              }
85            } //if(response)_end
86          }
87        }, // 第2个参数_end
88        new Response.ErrorListener() { //第3个参数，响应错误时的反馈信息
89          @Override
90          public void onErrorResponse(VolleyError error) {
91            Log.e("TAG 错误", error.getMessage(), error);
92          }
93        }
94      );  // 第64行开始的请求对象 StringRequest()_end
```

```
95        mQueue.add(stringRequest);  // 把请求对象 stringRequest 放置到队列中
96     }catch (Exception e){ }
97   } //getServerData()_end
98 } //class MainActivity_ _end
```

（6）修改配置文件，设置网络访问权限和支持 HTTP 访问权限（代码与例 7-4 相同，这里不再赘述）。

程序运行结果如图 7.12 所示。

图 7.12 访问远程数据库

习题 7

1. 设有 JSON 数组 [{"sid":1001, "name":"张大山"}, {"sid":1002, "name":"李小丽"}]，编写一个通过列表组件（ListView）显示 JSON 数组的程序。
2. 设计一个网络音乐播放器，通过点击列表中的歌曲名称播放所选取的音乐。
3. 进一步完善例 7-5 的程序，使之能显示数据库中查询到的多条数据。

第8章 数据存储

本章介绍 Android 系统的数据存储方法。Android 系统提供了多种数据存储的方式，有 SQLite 数据库存储方式、文件存储方式、XML 文件的 SharedPreferences 存储方式等。

8.1 内部存储空间和外部存储空间

Android 4.4 版本以后，系统的存储空间分为内部存储空间和外部存储空间，其中的外部存储空间可以通过插入 SD 卡（也称内存卡）来扩展容量，如图 8.1 所示。

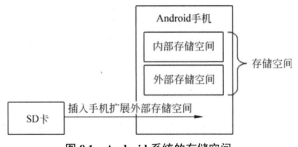

图 8.1 Android 系统的存储空间

1. 内部存储空间

内部存储空间的根目录为 data\data，存储应用程序所要保存数据的存放路径为 data\data\<应用程序包名>。应用 Device File Explorer 视图可以查看存放在内部存储空间的文件，如图 8.2 所示。

通常 SQLite 数据库、保存在应用程序资源中的文件及 SharedPreferences 中的数据，均存放在内部存储空间。

由于内部存储空间十分有限，且它也是 Android 系统本身和系统应用程序的数据存储所在地，一旦内部存储空间耗尽，Android 系统也就无法使用了。所以，对于内部存储空间要尽量节约使用。

2. 外部存储空间

外部存储空间的根目录为 storage\emulated\0，对于真实手机来说，若额外插入 SD 卡来

第8章　数据存储

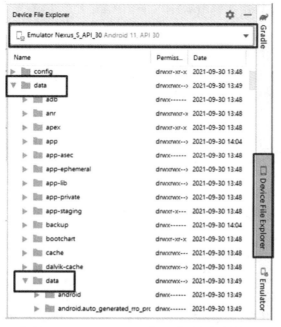

图 8.2　应用 Device File Explorer 视图查看内部存储空间的文件

扩展外部存储空间，则其 SD 卡的根目录为 storage\emulated\1。应用 Device File Explorer 视图可以查看存放在外部存储空间的文件，如图 8.3 所示。

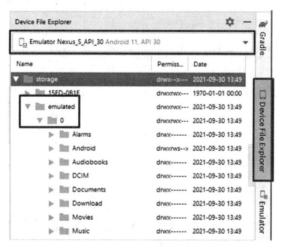

图 8.3　应用 Device File Explorer 视图查看外部存储空间的文件

从图 8.3 中可以看到，在外部存储空间的根目录 storage\emulated\0 下，有一些特定类型的子目录，可以通过 Environment.getExternalStoragePublicDirectory(String type)函数来获取这些目录，其中的参数 type 取值如下。

- DIRECTORY_ALARMS：警报铃声。
- DIRECTORY_DCIM：相机拍摄的图片和视频。
- DIRECTORY_DOWNLOADS：下载文件保存。
- DIRECTORY_MOVIES：电影的保存，比如通过 Google Play 下载的电影。

- DIRECTORY_MUSIC：音乐保存。
- DIRECTORY_NOTIFICATIONS：通知音乐保存。
- DIRECTORY_PICTURES：下载的图片。
- DIRECTORY_PODCASTS：用于保存 Podcast（博客）的音频文件。
- DIRECTORY_RINGTONES：保存铃声。

8.2 SQLite 数据库

8.2.1 SQLite 数据库简介

SQLite 数据库是一个关系型数据库，因为它很小，引擎本身只有一个大小不到 300KB 的文件，所以常作为嵌入式数据库内嵌在应用程序中。SQLite 生成的数据库文件是一个普通的磁盘文件，可以放置在任何目录下。SQLite 是用 C 语言开发的，开放源代码，支持跨平台，最大支持 2048GB 数据，并且被所有的主流编程语言支持。可以说 SQLite 是一个非常优秀的嵌入式数据库。

SQLite 数据库的管理工具很多，比较常用的有 SQLite Expert Professional，其功能强大，几乎可以在可视化的环境下完成所有数据库操作。用户使用它可以方便地创建数据表和对数据记录进行增加、删除、修改、查询操作。SQLite Expert Professional 的运行界面如图 8.4 所示。

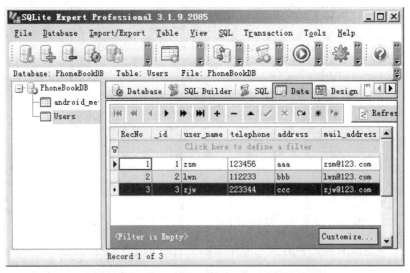

图 8.4　SQLite Expert Professional 的运行界面

在 Android 系统的内部集成了 SQLite 数据库，所以 Android 应用程序可以很方便地使用 SQLite 数据库来存储数据。

用户可以通过 Android 系统的 DDMS 工具将数据库文件复制到本地计算机上，应用 SQLite Expert Professional 对数据库进行操作，完成后再通过 DDMS 工具放回到设备中，当需要操作大量数据时是比较方便的方法。

8.2.2 管理和操作 SQLite 数据库的对象

Android 提供了创建和使用 SQLite 数据库的 API（Application Programming Interface，应用程序编程接口）。在 Android 系统中，主要由类 SQLiteDatabase 和 SQLiteOpenHelper 对 SQLite 数据库进行管理和操作，下面分别对它们进行介绍。

1. SQLiteDatabase 类

在 Android 中，主要由 SQLiteDatabase 对象对 SQLite 数据库进行管理，SQLiteDatabase 提供了一系列操作数据库的方法，可以对数据库进行创建、删除、执行 SQL 命令等操作，其常用的方法如表 8-1 所示。

表 8-1　SQLiteDatabase 的常用方法

方　　法	说　　明
openOrCreateDatabase(String path, SQLiteDatabase.CursorFactory factory)	打开或创建数据库
openDatabase(String path, SQLiteDatabase.CursorFactory factory, int flags)	打开指定的数据库
insert(String table, String nullColumnHack, ContentValues values)	新增一条记录
delete(String table,String whereClause, String[] whereArgs)	删除一条记录
query(String table, String[] columns, String selection, String[] selectionArgs, String groupBy, String having, String orderBy)	查询一条记录
update(String table, ContentValues values, String whereClause, String[] whereArgs)	修改一条记录
execSQL(String sql)	执行一条 SQL 语句
close()	关闭数据库

2. SQLiteOpenHelper 类

SQLiteOpenHelper 是 SQLiteDatabase 的一个辅助类，该类主要用于创建数据库，并对数据库的版本进行管理。当在程序中调用这个类的 getWritableDatabase()方法或者 getReadableDatabase()方法的时候，如果数据库不存在，那么 Android 系统就会自动创建一个数据库。

SQLiteOpenHelper 是一个抽象类，在使用时一般要定义一个继承 SQLiteOpenHelper 的子类，并实现其方法。SQLiteOpenHelper 的常用方法如表 8-2 所示。

表 8-2　SQLiteOpenHelper 的常用方法

方　　法	说　　明
onCreate(SQLiteDatabase)	首次生成数据库时调用该方法
onOpen(SQLiteDatabase)	调用已经打开的数据库
onUpgrade(SQLiteDatabase, int, int)	升级数据库时调用
getWritableDatabase()	以读/写方式创建或打开数据库
getReadableDatabase()	创建或打开数据库

8.2.3 SQLite 数据库的操作命令

对数据库的操作有 3 个层次，各层次的操作内容如下。
- 对数据库操作：建立数据库或删除数据库。
- 对数据表操作：建立、修改或删除数据库中的数据表。
- 对记录操作：对数据表中的数据记录进行添加、删除、修改、查询等操作。

下面按上述 3 个层次讲述对 SQLite 数据库操作命令的使用方法。

1. 创建及删除数据库

（1）创建数据库。

创建数据库的方法有多种，可以应用 SQLiteDatabase 对象 openDatabase()方法及 openOrCreateDatabase()方法创建数据库；也可以应用 SQLiteOpenHelper 的子类创建数据库；还可以应用 Activity 继承于父类 android.content.Context 创建数据库的方法 openOrCreateDatabase()来创建数据库。

Context 类的 openOrCreateDatabase（name, mode, factory）方法有以下 3 个参数。
第 1 个参数 name 为数据库名称。
第 2 个参数 mode 打开或创建数据库的模式，其模式有。
- MODE_PRIVATE：只可访问或调用模式，这是默认的模式。
- MODE_WORLD_READABLE：只读模式。
- MODE_WORLD_WRITEABLE：只写模式。

第 3 个参数 factory 为查询数据的游标，通常为 null。

例如，要创建一个名称为 PhoneBook.db 的数据库，其数据库的结构如下。

```
String TABLE_NAME = "Users";             //数据表名
String ID = "_id";                        //ID
String USER_NAME = "user_name";           //用户名
String ADDRESS = "address";               //地址
String TELEPHONE = "telephone";           //联系电话
String MAIL_ADDRESS = "mail_address";     //电子邮箱
```

则用 Activity 的 openOrCreateDatabase()方法创建数据库的代码如下。

```
SQLiteDatabase  db;
String db_name = "PhoneBook.db";
String sqlStr =
    "CREATE TABLE " + TABLE_NAME + " ("
        + ID + " INTEGER primary key autoincrement, "
        + USER_NAME + " text not null, "
        + TELEPHONE + " text not null, "
        + ADDRESS + " text not null, "
        + MAIL_ADDRESS + " text not null "+ ");";
int mode = Context.MODE_PRIVATE;
```

创建数据表的 SQL 语句

```
db = this.openOrCreateDatabase(Database_name, mode, null);    ← 创建数据库
db.execSQL(sqlStr);   ← 执行创建数据库的 SQL 语句
```

这时，通过 DDMS 可以看到，在 data\data\ xxxx（包名）\databases 路径下创建了数据库 PhoneBook.db。

（2）删除数据库。

当要删除一个指定的数据库文件时，需要应用 android.content.Context 类的 deleteDatabase(String name)方法来删除这个指定的数据库。

例如，要删除名为 PhoneBook.db 的数据库，则可以使用下列代码。

```
MainActivity.this.deleteDatabase("PhoneBook.db");
```

【例 8-1】 编写一个创建与删除数据库的演示程序。

视频讲解

（1）设计用户界面。在程序的用户界面中设置一个文本标签和两个按钮，如图 8.5 所示。

图 8.5 数据库测试程序的用户界面设计

（2）设计代码。

```
1   package com.example.ex8_1;
2   import androidx.appcompat.app.AppCompatActivity;
3   import android.os.Bundle;
4   import android.content.Context;
5   import android.database.sqlite.SQLiteDatabase;
6   import android.view.View;
7   import android.view.View.OnClickListener;
8   import android.widget.Button;
9   public class MainActivity extends Activity
10  {
11      Button creatBtn, deleteBtn;
12      DBCreate helper;
```

```java
13      SQLiteDatabase db1;
14      @Override
15      public void onCreate(Bundle savedInstanceState)
16      {
17          super.onCreate(savedInstanceState);
18          setContentView(R.layout.activity_main);
19          creatBtn=(Button)findViewById(R.id.creat1);
20          creatBtn.setOnClickListener(new mClick());
21          deleteBtn=(Button)findViewById(R.id.delete1);
22          deleteBtn.setOnClickListener(new mClick());
23      }
24      class mClick implements OnClickListener
25      {
26          @Override
27          public void onClick(View v)
28          {
29          if(v == creatBtn)
30              {
31                  DBCreate db =  new DBCreate();      ◄── 实例化创建数据库类
32              }
33            else if(v == deleteBtn)
34              {
35                  deleteDatabase(DBCreate.Database_name);   ◄── 删除数据库
36              }
37          }
38      }

39      class DBCreate
40      {
41        static final String Database_name = "PhoneBook.db";  ◄── 数据库名
42        private DBCreate()
43        {
44          SQLiteDatabase db;
45          String TABLE_NAME = "Users";          //数据表名
46          String ID = "_id";                    //ID
47          String USER_NAME = "user_name";       //用户名      ┐
48          String ADDRESS = "address";           //地址        ├── 定义数据库结构
49          String TELEPHONE = "telephone";       //联系电话    │
50          String MAIL_ADDRESS = "mail_address"; //电子邮箱    ┘
51          String DATABASE_CREATE =
52              "CREATE TABLE " + TABLE_NAME + " ("           ┐
53              + ID + " INTEGER primary key autoincrement,"  │
54              + USER_NAME + " text not null, "              ├── 创建数据表
55              + TELEPHONE + " text not null, "              │    的SQL语句
56              + ADDRESS + " text not null, "                │
57              + MAIL_ADDRESS + " text not null "+ ");";     ┘
58          int mode = Context.MODE_PRIVATE;
```

```
59            db = openOrCreateDatabase(Database_name, mode, null);    ← 创建数
60            db.execSQL(DATABASE_CREATE);    ← 执行创建数据库的 SQL 语句        据库
61        }
62    }
63 }
```

运行程序，单击"创建数据库"按钮，创建一个名为 PhoneBook.db 的数据库。该数据库保存在内部存储空间，通过 Device File Explorer 视图，在 data\data\<项目包名>\databases 路径下可以看到创建的数据库文件 PhoneBook.db，如图 8.6 所示。如果单击"删除数据库"按钮，数据库文件将被删除。

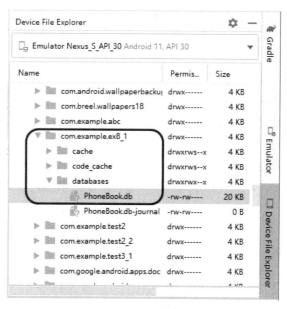

图 8.6 在 Device File Explorer 视图中查看创建的数据库文件 PhoneBook.db

2．数据表操作

（1）创建数据表。

创建数据表的步骤如下。

● 用 SQL 语句编写创建数据表的命令。

● 调用 SQLiteDatabase 的 execSQL()方法执行 SQL 语句。

（2）删除数据表。

删除数据表的步骤与创建数据表类似，先编写删除表的 SQL 语句，再调用 execSQL()方法执行 SQL 语句。

例如，要删除名为 Users 的数据表，则

```
String sql ="DROP TABLE Users";    ← 删除数据表 Users 的 SQL 语句
db.execSQL(sql);    ← 执行 SQL 语句
```

3. 数据记录操作

在数据表中把数据表的列称为字段，把数据表的每一行称为记录。对数据表中的数据进行操作处理主要是对其记录进行操作处理。

对数据记录的操作处理有两种方法，一种方法是编写一条对记录进行增、删、改、查的 SQL 语句，通过 exeSQL() 方法来执行；另一种方法是使用 Android 系统 SQLiteDatabase 对象的相应方法进行操作。对于使用 SQL 语句执行相应操作的办法，一般数据库的书籍均有介绍，这里不再赘述。下面仅介绍使用 SQLiteDatabase 对象操作数据记录的方法。

（1）新增记录。

新增记录的方法是使用 SQLiteDatabase 对象的 insert() 方法。

在 insert(String table, String nullColumnHack, ContentValues values) 方法中的 3 个参数的含义如下。

- 第 1 个参数 table：增加记录的数据表。
- 第 2 个参数 nullColumnHack：空列的默认值，通常为 null。
- 第 3 个参数 values：ContentValues 对象，即一个"键-值"对的字段名称，键名为表中的字段名，键值为要增加的记录数据值。通过 ContentValues 对象的 put() 方法把数据存放到 ContentValues 对象中。

例如，下列代码分别把 4 个"键-值"对数据存放到 values 对象中，其键名分别为 USER_NAME（用户名）、TELEPHONE（联系电话）、ADDRESS（住址）、MAIL_ADDRESS（电子邮箱）。

```
ContentValues values = new ContentValues();
values.put(USER_NAME, "张大山");
values.put(TELEPHONE, "13800012345");
values.put(ADDRESS, "南海仙人岛");
values.put(MAIL_ADDRESS, "abc@123.com");
db.insert("Users", null, values);
```
← 把数据存放到 values 中
← 将 values 数据添加到表 Users 中

（2）修改记录。

修改记录使用 SQLiteDatabase 对象的 update() 方法。在 update(String table, ContentValues values, String whereClause, String[] whereArgs) 方法中有 4 个参数，其含义如下。

- 第 1 个参数 table：修改记录的数据表。
- 第 2 个参数 values：ContentValues 对象，存放已做修改的数据的对象。
- 第 3 个参数 whereClause：修改数据的条件，相当于 SQL 语句的 where 子句。
- 第 4 个参数 whereArgs：修改数据值的数组。

例如，下列代码把用户名为张大山的经过修改的数据记录（替换数据表中原来的数据记录）存放到 values 中，其他数据值不变。

```
ContentValues values = new ContentValues();
values.put(TELEPHONE, "13811011011");
values.put(ADDRESS, "东海彭湖湾");
String where = "USER_NAME = 张大山";
db.update("Users", values, where ,null);
```
← 把修改的数据存放到 values 中
← values 数据替换表 Users 中满足条件的原数据

(3)删除记录。

删除记录使用 SQLiteDatabase 对象的 delete()方法。在 delete(String table,String whereClause, String[] whereArgs) 方法中有 3 个参数,其含义如下。
- 第 1 个参数 table:修改记录的数据表。
- 第 2 个参数 whereClause:删除数据的条件,相当于 SQL 语句的 where 子句。
- 第 3 个参数 whereArgs:删除条件的参数数组。

例如,下列代码把用户名为张大山的所有数据删除,即删除条件为"USER_NAME = 张大山"的记录。

```
String where = "USER_NAME = 张大山";
db.delete("Users", where ,null);    ← 删除表 Users 中满足条件的记录
```

(4)查询记录。

在数据库的操作命令中,查询数据的命令是最丰富、最复杂的。在 SQLite 数据库中使用 SQLiteDatabase 对象的 query()方法查询数据。在 query(String table, String[] columns, String selection, String[] selectionArgs, String groupBy, String having, String orderBy)方法中有 7 个参数,其含义如下。
- 第 1 个参数 table:查询记录的数据表。
- 第 2 个参数 columns:查询的字段,如为 null,则为所有字段。
- 第 3 个参数 selection:查询条件,可以使用通配符"?"。
- 第 4 个参数 selectionArgs:参数数组,用于替换查询条件中的"?"。
- 第 5 个参数 groupBy:查询结果按指定字段分组。
- 第 6 个参数 having:限定分组的条件。
- 第 7 个参数 orderBy:查询结果的排序条件。

(5)对查询结果 Cursor 的处理。

query()方法查询的数据均封装到查询结果 Cursor 对象之中,Cursor 相当于 SQL 语句中 resultSet 结果集上的一个游标,可以向前或向后移动。Cursor 对象的常用方法如下:
- moveToFirst():移动到第一行。
- moveToLast():移动到最后一行。
- moveToNext():向后移动一行。
- moveToPrevious():向前移动一行。
- moveToPosition(position):移动到指定位置。
- isBeforeFirst():判断是否指向第一条记录之前。
- isAfterLast():判断是否指向最后一条记录之后。

例如,查询用户表 Users 中的全部记录,并向前移动记录。

```
Cursor cursor;
cursor = db.query("Users", null , null, null, null, null, null);
cursor.moveToNext();
```

【例 8-2】 编写程序,建立一个通讯录,可以向前或向后浏览数据记录,也可以添加、修改、删除数据。

视频讲解

(1) 设计用户界面。

在界面设计中,用一个垂直线性布局嵌套了 3 个水平线性布局和一个表格布局,从而把整个界面划分为 4 部分。在第 1 个水平线性布局中嵌套了"建立数据库"和"打开数据库"按钮;在第 2 个水平线性布局中嵌套了"浏览通讯录"文本标签以及"上一记录"和"下一记录"按钮;在第 3 个水平线性布局中嵌套了"添加""修改""删除""关闭通讯录"按钮;在表格布局中设置表格为 4 行 2 列,每一行均包含一个文本标签和一个文本编辑框。

用户界面布局设计如图 8.7 所示。

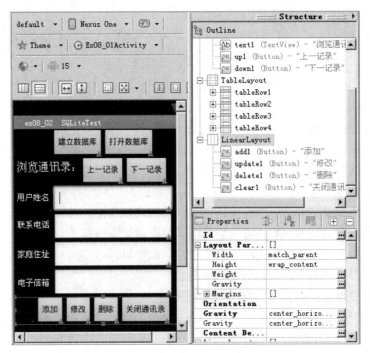

图 8.7 通讯录的界面布局设计

(2) 设计数据库程序 DBConnection.java。

```
1    package com.example.ex8_2;
2    import android.content.Context;
3    import android.database.Cursor;
4    import android.database.sqlite.SQLiteDatabase;
5    import android.database.sqlite.SQLiteOpenHelper;
6
7    public class DBConnection extends SQLiteOpenHelper
8    {
9      static final String Database_name = "PhoneBook.db";   ←定义数据库名
10     static final int Database_Version = 1;
11     SQLiteDatabase db;    ←定义 SQLiteDatabase 数据库对象
12     public int id_this;
13     Cursor cursor;
```

```
14      //定义数据库结构
15      static String TABLE_NAME = "Users";              //数据表名
16      static String ID = "_id";                         //ID
17      static String USER_NAME = "user_name";            //用户名
18      static String ADDRESS = "address";                //地址
19      static String TELEPHONE = "telephone";            //联系电话
20      static String MAIL_ADDRESS = "mail_address";      //电子邮箱
21      DBConnection(Context ctx)
22      {
23        super(ctx, Database_name, null, Database_Version);   ← 创建空数据库
24      }
25      public void onCreate(SQLiteDatabase database)
26      {
27        String sql = "CREATE TABLE " + TABLE_NAME + " ("
28          + ID + " INTEGER primary key autoincrement, "
29          + USER_NAME + " text not null, "                    ← 创建数据表
30          + TELEPHONE + " text not null, "
31          + ADDRESS + " text not null, "
32          + MAIL_ADDRESS + " text not null "+ ");";
33        database.execSQL(sql);
34      }
35      public void onUpgrade(SQLiteDatabase db, int oldVersion, int newVersion)
36      {         }
37    }
```

（3）设计控制程序 MainActivity.java。

```
1   package com.example.ex8_2;
2   import androidx.appcompat.app.AppCompatActivity;
3   import android.os.Bundle;
4   import android.content.ContentValues;
5   import android.content.Context;
6   import android.database.Cursor;
7   import android.database.sqlite.SQLiteDatabase;
8   import android.view.View;
9   import android.view.View.OnClickListener;
10  import android.widget.Button;
11  import android.widget.EditText;

12  public class MainActivity extends AppCompatActivity
13  {
14    static EditText mEditText01;
15    static EditText mEditText02;
16    static EditText mEditText03;
17    static EditText mEditText04;
18    Cursor cursor;
19    Button createBtn, openBtn, upBtn, downBtn;
```

```
20      Button addBtn, updateBtn, deleteBtn, closeBtn;
21      SQLiteDatabase db;
22      DBConnection helper;
23      public int id_this;
24      Bundle savedInstanceState;
25      //定义数据库结构
26      static String TABLE_NAME = "Users";              //数据表名
27      static String ID = "_id";                        //ID
28      static String USER_NAME = "user_name";           //用户名
29      static String ADDRESS = "address";               //地址
30      static String TELEPHONE = "telephone";           //联系电话
31      static String MAIL_ADDRESS = "mail_address";     //电子邮箱
32      @Override
33      public void onCreate(Bundle savedInstanceState)
34      {
35          super.onCreate(savedInstanceState);
36          setContentView(R.layout.main);
37          mEditText01 = (EditText)findViewById(R.id.EditText01);
38          mEditText02 = (EditText)findViewById(R.id.EditText02);
39          mEditText03 = (EditText)findViewById(R.id.EditText03);
40          mEditText04 = (EditText)findViewById(R.id.EditText04);
41          createBtn = (Button)findViewById(R.id.createDatabase1);
42          createBtn.setOnClickListener(new ClickEvent());
43          openBtn = (Button)findViewById(R.id.openDatabase1);
44          openBtn.setOnClickListener(new ClickEvent());
45          upBtn=(Button)findViewById(R.id.up1);
46          upBtn.setOnClickListener(new ClickEvent());
47          downBtn=(Button)findViewById(R.id.down1);
48          downBtn.setOnClickListener(new ClickEvent());
49          addBtn = (Button)findViewById(R.id.add1);
50          addBtn.setOnClickListener(new ClickEvent());
51          updateBtn = (Button)findViewById(R.id.update1);
52          updateBtn.setOnClickListener(new ClickEvent());
53          deleteBtn = (Button)findViewById(R.id.delete1);
54          deleteBtn.setOnClickListener(new ClickEvent());
55          closeBtn = (Button)findViewById(R.id.clear1);
56          closeBtn.setOnClickListener(new ClickEvent());
57      }
58      class ClickEvent implements OnClickListener
59      {
60        public void onClick(View v)
61        {
62          switch(v.getId())
63          {
64            case R.id.createDatabase1:
65              helper = new DBConnection(MainActivity.this);
66              SQLiteDatabase db=helper.getWritableDatabase();
67              break;
```

```
68          case R.id.openDatabase1:
69              db = openOrCreateDatabase("PhoneBook.db",
70                      Context.MODE_PRIVATE, null) ;
71              cursor = db.query("Users",
72                      null , null, null, null, null, null);
73              cursor.moveToNext();
74              upBtn.setClickable(true);
75              downBtn.setClickable(true);
76              deleteBtn.setClickable(true);
77              updateBtn.setClickable(true);
78              break;
79          case  R.id.up1:
80              if(!cursor.isFirst())
81                  cursor.moveToPrevious();
82              datashow();
83              break;

84          case R.id.down1:
85              if(!cursor.isLast())
86                  cursor.moveToNext();
87              datashow();
88              break; //
89          case R.id.add1:
90              add();
91              onCreate(savedInstanceState);
92              break;
93          case R.id.update1:
94              update();
95              onCreate(savedInstanceState);
96              break;
97          case  R.id.delete1:
98              delete();
99              onCreate(savedInstanceState);
100             break;
101         case R.id.clear1:
102             cursor.close();
103             mEditText01.setText("数据库已关闭");
104             mEditText02.setText("数据库已关闭");
105             mEditText03.setText("数据库已关闭");
106             mEditText04.setText("数据库已关闭");
107             upBtn.setClickable(false);
108             downBtn.setClickable(false);
109             deleteBtn.setClickable(false);
110             updateBtn.setClickable(false);
111             break;
112         }
113     }
```

注释说明:
- 68–78行: 查询Users数据表
- 79–83行: 按下"上一记录"按钮，向前查询
- 84–88行: 按下"下一记录"按钮，向后查询
- 89–92行: 按下"添加"按钮，新增一行数据
- 93–96行: 按下"修改"按钮，更新一行数据
- 97–100行: 按下"删除"按钮，删除一行数据
- 101行: 按下"关闭通讯录"按钮，关闭数据库
- 107–110行: 使按钮不可用

```
114    }
115    /* 显示记录 */
116    void datashow()
117    {
118      id_this = Integer.parseInt(cursor.getString(0));
119      String user_name_this = cursor.getString(1);
120      String telephone_this = cursor.getString(2);
121      String address_this = cursor.getString(3);
122      String mail_address_this = cursor.getString(4);
123      mEditText01.setText(user_name_this);
124      mEditText02.setText(telephone_this);
125      mEditText03.setText(address_this);
126      mEditText04.setText(mail_address_this);
127    }
128    /* 添加记录 */
129    void add()
130    {
131      ContentValues values1 = new ContentValues();
132      values1.put(USER_NAME, MainActivity.mEditText01.getText().toString());
133      values1.put(TELEPHONE, MainActivity.mEditText02.getText().toString());
134      values1.put(ADDRESS, MainActivity.mEditText03.getText().toString());
135      values1.put(MAIL_ADDRESS,
136              MainActivity.mEditText04.getText().toString());
137      SQLiteDatabase db2 = helper.getWritableDatabase();
138      db2.insert(TABLE_NAME, null, values1);       ← 将数据插入数据表
139      db2.close();
140    }
141    /* 修改记录 */
142    void update()
143    {
144      ContentValues values = new ContentValues();
145      values.put(USER_NAME, MainActivity.mEditText01.getText().toString());
146      values.put(TELEPHONE, MainActivity.mEditText02.getText().toString());
147      values.put(ADDRESS, MainActivity.mEditText03.getText().toString());
148      values.put(MAIL_ADDRESS,
149              MainActivity.mEditText04.getText().toString());
150      String where1 = ID + " = " + id_this;
151      SQLiteDatabase db1 = helper.getWritableDatabase();
152      db1.update(TABLE_NAME, values, where1 ,null);   ← 替换数据表中的原数据
153      db1.close();
154    }
155    /* 删除记录 */
156    void delete()
157    {
158      String where = ID + " = " + id_this;
159      db.delete(TABLE_NAME, where ,null);     ← 删除数据表中的数据
160      db = helper.getWritableDatabase();
```

注：118—122 行 读取各字段数据

```
161        db.close();
162    }
163 }
```

程序运行结果如图 8.8 所示。

图 8.8　通讯录运行示例

8.3 文件处理

8.3.1 输入/输出流

程序可以理解为数据输入、输出以及数据处理的过程。在程序执行过程中，通常需要读取处理数据，并且将处理后的结果保存起来。Android 系统提供了对数据流进行输入与输出的方法。

1. 文件与目录管理的 File 类

Android 系统处理文件时直接调用 Java 语言的 java.io 包中的 File 类。每个 File 类的对象都对应了系统的一个文件或目录，所以在创建 File 类对象时需要指明它所对应的文件或目录名。

下面是应用 File 类创建目录对象及文件对象的示例。

假设：

```
String sdir = "data/jtest";
String sfile = "FileIO.data";
```

则：

```
File Fdir = new File ( sdir );           //目录对象
File Ffile = new File ( Fdir, sfile );   //文件对象
```

一个对应于某文件或目录的 File 对象一经创建就可以通过调用它的方法来获得文件或目录的属性。在建立文件类的一个实例后，可以查询这个文件对象，用测试方法获得有关文件或目录的有关信息，如检测文件和目录的属性。File 类的常用方法如表 8-3 所示。

表 8-3　File 类的常用方法

方　　法	说　　明
exists( )	判断文件或目录是否存在
isFile( )	判断对象是否为文件
isDirectory( )	判断对象是否为目录
getName( )	返回文件名或目录名
getPath( )	返回文件或目录的路径
length( )	返回文件的字节数
renameTo(File newFile)	将文件重命名成 newFile 对应的文件名
delete( )	将当前文件删除
mkdir( )	创建当前目录的子目录

2．文件输入/输出流

在 Android 中处理二进制文件使用字节输入/输出流，处理字符文件使用字符输入/输出流。下面介绍对文件进行输入/输出处理的 4 个类。

- FileInputStream：字节文件输入流。
- FileOutputStream：字节文件输出流。
- FileReader：字符文件输入流。
- FileWriter：字符文件输出流。

8.3.2　处理文件流

1．用文件输出流保存文件

（1）FileOutputStream 类。

FileOutputStream 类是从 OutputStream 类派生出来的输出类，它具有向文件中写数据的能力。它的构造方法有以下 3 种形式。

- FileOutputStream(String filename)。
- FileOutputStream(File file)。
- FileOutputStream(FileDescriptor fdObj)。

其中各参数的含义如下。

- String filenam：指定的文件名，包括路径。
- File file：指定的文件对象。
- FileDescriptor fdObj：指定的文件描述符。

用户也可以通过 Context.openFileOutput()方法获取 FileOutputStream 对象。

（2）把字节发送到文件输出流的 write()方法。

输出流只是建立了一条通往数据要去的目的地的通道,数据并不会自动进入输出流通道,用户要使用文件输出流的 write()方法把字节发送到输出流。

使用 write()方法有以下 3 种格式。

- write(int b):将指定字节写入此文件输出流。
- write(byte[] b):将 b.length 个字节从指定字节数组写入此文件输出流。
- write(byte[] b, int off, int len):将指定字节数组中从偏移量 off 开始的 len 字节写入此文件输出流。

【例 8-3】 把字符串"Hello World!"保存到本地资源的 test.txt 文件中。

在项目设计中设置一个"保存文件"按钮,其控制程序的代码如下。

```
1    package com.example.ex8_3;
2    import androidx.appcompat.app.AppCompatActivity;
3    import android.os.Bundle;
4    import android.content.Context;
5    import android.view.View;
6    import android.widget.Button;
7    import java.io.FileNotFoundException;
8    import java.io.FileOutputStream;
9    import java.io.IOException;

10   public class MainActivity extends AppCompatActivity {
11     Button saveBtn;
12     @Override
13     protected void onCreate(Bundle savedInstanceState) {
14       super.onCreate(savedInstanceState);
15       setContentView(R.layout.activity_main);
16       saveBtn=(Button)findViewById(R.id.button);
17       saveBtn.setOnClickListener(new mClick());
18     }
19     class mClick implements View.OnClickListener{
20       @Override
21       public void onClick(View v) {
22         savefile();
23       }
24     }
25     void savefile()
26     {
27       String fileName="test.txt";
28       String str = "Hello World!";
29       FileOutputStream f_out;
30       try {
31         f_out = openFileOutput(fileName, Context.MODE_PRIVATE);    ← 文件输出流
32         f_out.write(str.getBytes());    ← 写操作
33       }
34       catch (FileNotFoundException e) {e.printStackTrace();}
```

```
35        catch (IOException e) {e.printStackTrace();}
36    }
37 }
```

则文件 test.txt 保存在 data\data\<包名>\files\目录下，应用 Device File Explorer 视图可以查看到保存在本地资源目录下的文件，如图 8.9 所示。

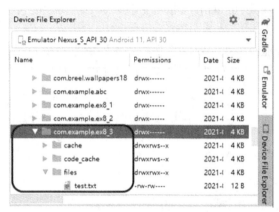

图 8.9　应用 Device File Explorer 视图查看保存到本地资源目录下的文件

2．用文件输入流读取文件

（1）FileInputStream 类。

FileInputStream 类是从 InputStream 类中派生出来的输入流类，它用于处理二进制文件的输入操作。它的构造方法有以下 3 种形式。

- FileInputStream(String filename)。
- FileInputStream(File file)。
- FileInputStream(FileDescriptor fdObj)。

其各参数的含义和 FileOutputStream 一样。

用户也可以通过 Context.openFileInput()方法获取 FileInputStream 对象。

（2）从文件输入流中读取字节的 read()方法。

文件输入流只是建立了一条通道，应用程序可以通过这个通道读取数据，而要实现读取数据的操作，需要使用 read()方法。

使用 read()方法有以下 3 种格式。

- int read()。
- int read(byte b[ ])。
- int read(byte b[ ],int off, int len)。

第 1 种格式每次只能从输入流中读取 1 字节的数据。该方法返回的是一个 0～255 的整数值，若为文本类型的数据返回的是 ASCII 值。如果该方法到达输入流的末尾则返回-1。

第 2 种格式和第 3 种格式以字节型数组作为参数，一次可以读取多字节，读入的字节数据直接放入字节数组 b 中，并返回实际读取的字节数。如果该方法到达输入流的末尾则返回-1。

第 3 种格式设置了偏移量（off）。这里的偏移量是指可以从字节型数组的第 off 个位置起,读取 len 字节数据。

【例 8-4】 读取本地资源文件 test.txt 中的内容。

在项目设计中设置一个"读取资源文件"按钮,其按钮事件调用方法如下。

```
1    void readfile()
2    {
3      String fileName="test.txt", str ;
4      byte[] buffer = new byte[1024];      ← 设 1 字节数组,用于存放读取的数据
5      FileInputStream in_file=null;
6      try {
7         in_file = openFileInput(fileName);  ← 文件输入流
8         int  bytes = in_file.read(buffer);
9         str = new String(buffer, 0, bytes);  ← 将读取到的文件数据转换成字符串
10        Toast.makeText(MainActivity.this,
11               "文件内容: " + str, Toast.LENGTH_LONG).show();
12     }
13     catch (FileNotFoundException e) { System.out.print("文件不存在");}
14     catch (IOException e) { System.out.print("IO 流错误");
15   }
```

3. 对外部存储空间文件的读/写

上述应用文件流对文件的读/写操作也适用于外部存储空间,但在处理上稍有不同,因为这里要考虑对外部存储空间文件的读/写权限。

(1) 环境变量访问类 Environment。

Environment 为提供的环境变量访问类,在 Android 程序中对外部存储空间的文件进行读/写操作时,经常需要应用它的以下两个方法。

● getExternalStorageState():获取当前存储设备状态。

● getExternalStoragePublicDirectory(String type):获取外部存储空间目录。

(其中,参数 type 为目录类型,见 8.1 节内部存储空间和外部存储空间)

(2) 外部存储空间文件的读/写权限。

在应用程序配置文件 AndroidManifest.xml 中,要加入允许对外部存储空间文件进行读/写操作的权限语句。

● 允许对外部存储空间文件写入数据的权限语句:

```
<uses-permission
    android:name="android.permission.WRITE_EXTERNAL_STORAGE">
</uses-permission>
```

● 允许对外部存储空间文件读取数据的权限语句:

```
<uses-permission
    android:name="android.permission. READ_EXTERNAL_STORAGE">
</uses-permission>
```

【例 8-5】 外部存储空间文件的读/写示例。

(1) 新建应用程序,设置一个文本编辑框和两个按钮,按钮分别为"保存到外部存储"和"读取外部存储文件"。

(2) 在 AndroidManifest.xml 文件中,加入允许对外部存储空间文件进行读/写数据操作权限的语句,其代码如下。

```xml
1   <?xml version="1.0" encoding="utf-8"?>
2   <manifest xmlns:android="http://schemas.android.com/apk/res/android"
3       package="com.example.ex8_5_sdcard">
4       <application
5           android:allowBackup="true"
6           android:icon="@mipmap/ic_launcher"
7           android:label="@string/app_name"
8           android:roundIcon="@mipmap/ic_launcher_round"
9           android:supportsRtl="true"
10          android:theme="@style/Theme.ex8_5_sdcard">
11          <activity android:name=".MainActivity">
12              <intent-filter>
13                  <action android:name="android.intent.action.MAIN" />
14                  <category android:name="android.intent.category.LAUNCHER" />
15              </intent-filter>
16          </activity>
17      </application>
18      <!--外部存储的读写权限-->
19      <uses-permission android:name="android.permission.READ_EXTERNAL_STORAGE"/>
20      <uses-permission android:name="android.permission.WRITE_EXTERNAL_STORAGE"/>
21  </manifest>
```

(3) 编写控制程序 MainActivity.java。

```java
1   package com.example.ex8_5;
2   import androidx.appcompat.app.AppCompatActivity;
3   import android.os.Bundle;
4   import android.os.Environment;
5   import android.util.Log;
6   import android.view.View;
7   import android.view.View.OnClickListener;
8   import android.widget.Button;
9   import android.widget.EditText;
10  import android.widget.Toast;
11  import java.io.File;
12  import java.io.FileInputStream;
13  import java.io.FileNotFoundException;
14  import java.io.FileOutputStream;
15  import java.io.IOException;
16  public class MainActivity extends Activity
17  {
```

```
18    Button savesdBtn, readsdBtn;
19    EditText edit;
20    String fileName="test.txt";
21    String str="";
22    @Override
23    public void onCreate(Bundle savedInstanceState)
24    {
25        super.onCreate(savedInstanceState);
26        setContentView(R.layout.activity_main);
27        edit=(EditText)findViewById(R.id.edit1);
28        readBtn.setOnClickListener(new mClick());
29        savesdBtn=(Button)findViewById(R.id.button3);
30        savesdBtn.setOnClickListener(new mClick());
31        readsdBtn=(Button)findViewById(R.id.button4);
32        readsdBtn.setOnClickListener(new mClick());
33    }
34    //按钮事件
35    class mClick implements OnClickListener
36    {
37        @Override
38        public void onClick(View v)
39        {
40         if(v == savesdBtn)
41         {
42             saveSDcar();          ◄── 写入外部存储空间
43         }
44         else if(v == readsdBtn)
45         {
46             readsdcard(fileName);  ◄── 读取外部存储空间文件数据
47         }
48        }
49    }
50    //保存文件到外部存储空间
51    void saveSDcar()
52    {
53      str= edit.getText().toString();
54      //获取外部存储路径
55      String type = Environment.DIRECTORY_DOWNLOADS;
56      File fileDir = Environment.getExternalStoragePublicDirectory(type);
57      File sdfile = new File(fileDir + "/" + fileName);
58      try {
59          FileOutputStream f_out = new FileOutputStream(sdfile);
60          f_out.write(str.getBytes());
61          f_out.close();
62          Toast.makeText(MainActivity.this, "文件保存到内存卡路径: "
63                  + sdfile.toString(), Toast.LENGTH_LONG).show();
64      }catch (FileNotFoundException e) {
65          Log.e("TAG", "(1)出错了，SD卡文件创建失败");
66          e.printStackTrace();
```

```
67          }catch (IOException e) {
68              Log.e("TAG", "（2）出错了，文件没有保存到 SD 卡");
69              e.printStackTrace();
70          }
71      }
72      //从外部存储空间读取文件内容
73      void readsdcard(String fileName)
74      {
75          if(Environment.getExternalStorageState()         ← 判断是否允许读/写操作
76              .equals(Environment.MEDIA_MOUNTED))
77          {
78            File path = Environment                        ← 获取存储空间路径
79                  .getExternalStorageDirectory();
80            File sdfile= new File(path, "Download/" + fileName);
81            Log.v("Path:", sdfile.toString());
82            try {
83                FileInputStream in_file =new FileInputStream(sdfile);
84                byte[]  buffer = new byte[1024];
85                int bytes = in_file.read(buffer);           ← 读取文件数据到字节数组
86                str = new String(buffer, 0, bytes);
87                edit.setText(str);
88                Log.v("test.txt:", str);
89            } catch (FileNotFoundException e){
90                System.out.print("文件不存在");
91            }
92            catch (IOException e) { System.out.print("IO 流错误");}
93          }
94      }
95  }
```

运行程序，在输入框中输入文字信息，单击"保存文件"按钮后，应用 Device File Explorer 视图可以查看到保存在 storage\emulated\0\Download 目录下的 test.txt 文件，结果如图 8.10 所示。

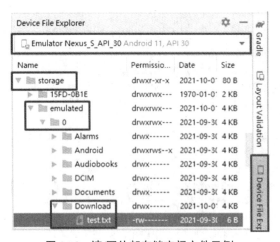

图 8.10　读/写外部存储空间文件示例

8.4 轻量级存储 SharedPreferences

Android 系统提供了一个存储少量数据的轻量级的数据存储方式 SharedPreferences。该存储方式类似于 Web 程序中的 Cookie，通常用它来保存一些配置文件数据、用户名及密码等。SharedPreferences 采用"键-值"对的形式组织和管理数据，其数据存储在 XML 格式的文件中。

使用 SharedPreferences 方式存储数据需要用到 SharedPreferences 和 SharedPreferences.Editor 接口，这两个接口在 android.content 包中。

SharedPreferences 对象由 Context.getSharedPreferences (String name, int mode)方法构造，它有两个参数，其含义如下。

（1）第 1 个参数 name 为保存数据的文件名，因为 SharedPreferences 使用 XML 文件保存数据，getSharedPreferences(name,mode)方法的第 1 个参数用于指定该文件的名称，名称不用带后缀，后缀会由 Android 自动加上。该 XML 文件存放在 data\data\<包名>\shared_prefs 目录下。

（2）第 2 个参数 mode 为操作模式。

- MODE_PRIVATE：这是默认的形式，配置文件只允许本程序和享有本程序 ID 的程序的访问。
- MODE_WORLD_READABLE：允许其他的应用程序读文件。
- MODE_WORLD_WRITEABLE：允许其他的应用程序写文件。
- MODE_MULTI_PROCESS：主要用于多任务，当多个进程共同访问时，必须指定这个参数。

SharedPreferences 接口的常用方法如表 8-4 所示。

表 8-4 SharedPreferences 接口的常用方法

方 法	说 明
edit()	建立一个 SharedPreferences.Editor 对象
contains(String key)	判断是否包含该键值
getAll()	返回所有配置信息
getBoolean(String key,Boolean defValue)	获得一个 boolean 类型数据
getFloat(String key,float defValue)	获得一个 float 类型数据
getInt(String key,int defValue)	获得一个 int 类型数据
getLong(String key,long defValue)	获得一个 long 类型数据
getString(String key,String defValue)	获得一个 string 类型数据

SharedPreferences.Editor 接口用于存储 SharedPreferences 对象的数据值，其常用方法如表 8-5 所示。

SharedPreferences.Editor 的 putXXX 方法以"键-值"对的形式存储数据，最后一定要调用 commit 方法提交数据，文件才能保存。

表 8-5　SharedPreferences.Editor 接口的常用方法

方　　法	说　　明
clear()	清除所有数据值
commit()	保存数据
putBoolean(String key, boolean value)	保存一个 boolean 类型数据
putFloat(String key, float value)	保存一个 float 类型数据
putInt(String key, int value)	保存一个 int 类型数据
putLong(String key, long value)	保存一个 long 类型数据
putString(String key, String value)	保存一个 string 类型数据
remove(String key)	删除键名 key 所对应的数据值

读取数据非常简单，直接调用 SharedPreferences 对象相应的 getXXX 方法即可获得数据。

【例 8-6】应用 SharedPreferences 对象将一个客户的联系电话保存到电话簿中。

设客户名为 zsm，其电话为 123456。故设电话簿的文件名为 phoneBook，其数据的"键-值"对为（"name", "zsm"）和（"phone", "123456"）。其主程序设计如下。

```
1   package com.example.ex8_6;
2   import androidx.appcompat.app.AppCompatActivity;
3   import android.os.Bundle;
4   import android.content.Context;
5   import android.content.SharedPreferences;
6   import android.view.View;
7   import android.view.View.OnClickListener;
8   import android.widget.Button;
9   import android.widget.Toast;
10  public class MainActivity extends AppCompatActivity
11  {
12      SharedPreferences settings;
13      Button saveBtn;
14      @Override
15      public void onCreate(Bundle savedInstanceState)
16      {
17          super.onCreate(savedInstanceState);
18          setContentView(R.layout.activity_main);
19          saveBtn=(Button)findViewById(R.id.button1);
20          saveBtn.setOnClickListener(new mClick());
21      }
22      //按钮事件
23      class mClick implements OnClickListener
24      {
25          public void onClick(View arg0)
26          {
27              settings = getSharedPreferences("phoneBook", Context.MODE_PRIVATE);
```

```
28        SharedPreferences.Editor editor = settings.edit();
29        editor.putString("name", "zsm");
30        editor.putString("phone", "123456");
31        editor.commit();
32        Toast.makeText(MainActivity.this, "保存成功！",
33                                         Toast.LENGTH_LONG).show();
34    }
35  }
36 }
```

运行程序，单击"保存文件"按钮后，生成的数据文件 phoneBook.xml 保存在内部存储空间 data\data\com.example.ex8_6\shared_prefs\目录下，扩展名.xml 由系统自动生成。应用 Device File Explorer 视图可以查看到该文件，如图 8.11 所示。

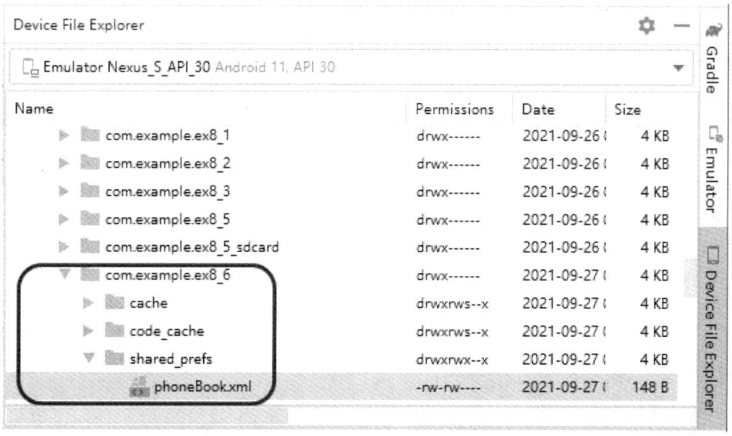

图 8.11 应用 Device File Explorer 视图查看保存的 XML 文件

右击 Device File Explorer 视图的 phoneBook.xml 项，选择 Open 命令，可以打开该数据文件，其内容如下。

```
<?xml version='1.0' encoding='utf-8' standalone='yes' ?>
<map>
      <string name="phone">123456</string>
      <string name="name">zsm</string>
</map>
```

习题 8

1. 编写一个小型商场的销售管理系统，可以输入商品的名称、数量、单价，并具有汇总功能。

2. 设计一个如图 8.12 所示的简易记事本，以文件形式保存。

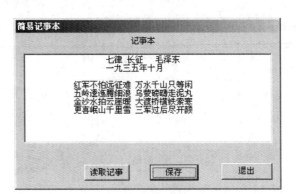

图 8.12　简易记事本

第9章 OpenCV应用实战——人脸美颜与人脸检测

9.1 OpenCV 图像处理

OpenCV 的全称是源代码开放的计算机视觉库（Open Source Computer Vision Library），它是由 Intel 公司基于 C/C++编写的。OpenCV 可以应用于人机互动、物体识别、图像分割、人脸识别、运动跟踪、机器人、机器视觉、汽车自动驾驶等众多领域。

9.1.1 搭建 OpenCV Android 开发环境

视频讲解

1. 下载 OpenCV Android SDK

OpenCV Android SDK 包的下载地址为 https://opencv.org/releases。
选择相应版本中的 Android 选项，进入 OpenCV Android 的下载页面，如图 9.1 所示。

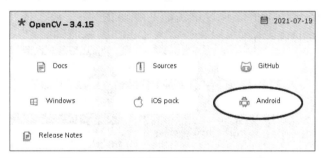

图 9.1 选择相应版本中的 Android 选项

2. 安装开发包

下载的 OpenCV Android SDK 是一个压缩包，只需要解压即可，其解压后的目录结构如图 9.2 所示。

3. 搭建开发环境

（1）导入模块。

新建一个空白应用项目，选择菜单 File→New→Import Module 命令，在弹出的 Import

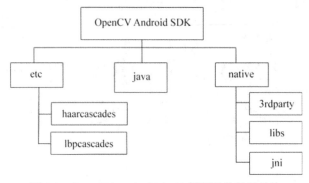

图 9.2　OpenCV Android SDK 解压后的目录结构

module from source 对话框中选择 OpenCV-android-sdk 解压目录下的 sdk\java，如图 9.3 所示。

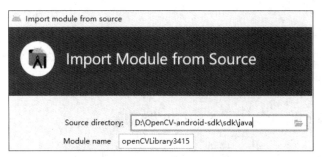

图 9.3　导入 OpenCV Android SDK 模块

（2）建立应用程序与模块的依赖关系。

选择菜单 File→Project Structure 命令，在弹出的 Project Structure 对话框中选择 Dependencies→Modules→app 选项，单击右边栏中的"＋"，选择其中的 Add Module Dependency 选项，如图 9.4 所示。

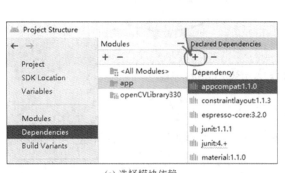

(a) 选择模块依赖　　　　　　　　　　　(b) 添加模块依赖

图 9.4　建立应用程序与模块的依赖关系

（3）复制 libs 文件夹并更名为 jniLibs。

把 OpenCV-android-sdk\sdk\native 下的 libs 文件夹复制到应用项目 app\src\main 下，并

将 libs 改名为 jniLibs，如图 9.5 所示。

(a) native 下的 libs 文件夹　　(b) 复制后的 jniLibs 文件夹

图 9.5　将 sdk\native 下的 libs 文件夹复制到\app\src\main 下，并改名为 jniLibs

（4）添加 jniLibs 路径到 build.gradle 中。

打开 app 目录下的 build.gradle 文件，在 build.gradle 的 android 节点中加入自定义 jniLibs 的路径。

```
sourceSets {
    main {
        jniLibs.srcDirs = ['src/main/jniLibs']
    }
}
```

修改后的 build.gradle 文件如图 9.6 所示。

图 9.6　修改后的 build.gradle 文件

9.1.2　Mat 对象和 Bitmap 对象

1. Mat 对象

Mat 是 OpenCV 中用来存储图像像素数据和图像宽、高、类型、维度等基本信息的内存对象。在 OpenCV 对图像进行处理时，通常都是通过 Mat 对象来处理的。

2. Bitmap 对象

Bitmap 是一个与 Mat 相似的对象，通过它可以获取图像的像素数据和常见属性，还可以通过修改 Bitmap 对象图像的数据来呈现不同的图像显示效果等。

假设应用程序资源中有一个图片文件 img.jpg，将其加载成 Bitmap 对象的方法如下。

```
Bitmap bitmap = BitmapFactory.decodeResource(
            MainActivity.this.getResources(),
            R.drawable.img);
```

3. Mat 对象与 Bitmap 对象的转换

Mat 对象与 Bitmap 对象可以相互转换，其转换方法如下。

（1）Mat 对象转换成 Bitmap 对象。

```
matToBitmap(dst, bitmap);
```

其中，dst 是 Mat 对象，bitmap 是 Bitmap 对象。

（2）Bitmap 对象转换成 Mat 对象。

```
bitmapToMat(bitmap, dst);
```

视频讲解

【例 9-1】 应用 OpenCV 将彩色图片转换成灰度图片。

（1）创建应用程序框架，设置 OpenCV 的环境，并将一张事先准备好的彩色图片 img2.jpg 复制到 res\drawable 目录下。

（2）设计界面布局。

在界面中设置一个显示图片的 ImageView 组件和两个按钮组件，一个按钮用于将彩色图片转换成灰度图片，另一个按钮用于还原彩色图片。

（3）设计控制程序代码。

将彩色图片转换成灰度图片的主要步骤如下。

① 将 Bitmap 对象转换成 Mat 对象 src。

```
Utils.bitmapToMat(bitmap_gray, src);
```

② 将彩色的 Mat 对象 src 转换成灰度的 Mat 对象 dst。

```
Imgproc.cvtColor(src, dst, Imgproc.COLOR_BGRA2GRAY);
```

③ 再将灰度 Mat 对象 dst 转换成 Bitmap 对象，用于显示灰度图像。

```
Utils.matToBitmap(dst, bitmap_gray);
```

完整的控制程序代码如下。

```
1    package com.example.ex9_1;
2    import androidx.appcompat.app.AppCompatActivity;
3    import android.os.Bundle;
4    import android.graphics.Bitmap;
```

第9章 OpenCV应用实战——人脸美颜与人脸检测

```java
5   import android.graphics.BitmapFactory;
6   import android.view.View;
7   import android.widget.Button;
8   import android.widget.ImageView;
9   import android.widget.TextView;
10  import org.opencv.android.OpenCVLoader;
11  import org.opencv.android.Utils;
12  import org.opencv.core.Mat;
13  import org.opencv.imgproc.Imgproc;

14  public class MainActivity extends AppCompatActivity {
15  TextView txt;
16  Button btn1, btn2;
17  Bitmap bitmap;
18  ImageView imageView;
19  @Override
20  protected void onCreate(Bundle savedInstanceState) {
21      super.onCreate(savedInstanceState);
22      setContentView(R.layout.activity_main);
23      txt = (TextView)findViewById(R.id.textView);
24      imageView = (ImageView) findViewById(R.id.imageView);
25      btn1 = (Button)findViewById(R.id.button);
26      btn1.setOnClickListener(new mClick1());
27      btn2 = (Button)findViewById(R.id.button2);
28      btn2.setOnClickListener(new mClick2());
29  }
30  public Bitmap getBitmap(){  //将图片加载为Bitmap对象
31      bitmap = BitmapFactory.decodeResource(
32          MainActivity.this.getResources(),
33          R.drawable.img2);
34      return bitmap;
35  }
36  private void iniLoadOpenCV(){  //加载OpenCV
37      boolean success = OpenCVLoader.initDebug();
38      if(success){
39          txt.setText("openCV loaded ...");
40      } else{ txt.setText("openCV not load!!! ");}
41  }
42  class mClick1 implements View.OnClickListener{
43      @Override
44      public void onClick(View v) {
45          iniLoadOpenCV();
46          Mat src = new Mat();
47          Mat dst = new Mat();
48          Bitmap bitmap_gray= getBitmap();
49          Utils.bitmapToMat(bitmap_gray, src);
50          Imgproc.cvtColor(src, dst, Imgproc.COLOR_BGRA2GRAY);
```

```
51              Utils.matToBitmap(dst, bitmap_gray);
52              imageView.setImageBitmap(bitmap_gray);
53              src.release();
54              dst.release();
55          }
56      }
57      class mClick2 implements View.OnClickListener{
58          @Override
59          public void onClick(View v) {
60              Bitmap bitmap= getBitmap();
61              imageView.setImageBitmap(bitmap);
62          }
63      }
64  }
```

程序运行结果如图 9.7 所示。

图 9.7 彩色图片转换成灰度图片

视频讲解

9.1.3 图像的模糊与锐化

1. 图像模糊的原理

图像的模糊算法是一种处理图像的数据平滑技术。所谓"模糊",可以理解成每一个像素都取周边像素的平均值。

例如,设有一图像,其位于中间点的值是 2,周边点的值都是 1,现在让中间点取周围点的平均值,则其值就会变成 1,如图 9.8 所示。

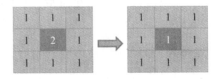

图 9.8 图像的像素点

在数值上,这是一种"平滑化"。在图形显示上,中间点失去细节,就会产生模糊的效果。

用来对图像矩阵进行平滑的矩阵称为卷积核，定义3×3模糊卷积核代码如下。

```
Mat kernel = new Mat(3, 3, CvType.CV_32FC1);
float[] data = new float[]{0,          1.0f/8.0f,       0,
                           1.0f/8.0f,  0.5f,            1.0f/8.0f,
                           0,          1.0f/8.0f,       0};
kernel.put(0, 0, data);
```

2．图像锐化处理

图像锐化处理是对图像模糊的逆运算，它是通过增强高频分量来减少图像中的模糊，因此又称为高通滤波（High Pass Filter）。利用图像处理的锐化算子可以使图像的边缘变得清晰，两种常见的锐化算子如图9.9所示。

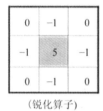
（锐化算子）

（强化锐化算子）

图9.9　两种常见的锐化算子

自定义实现锐化算子的代码如下。

```
Mat kernel = new Mat(3, 3, CvType.CV_32FC1);
float[] data = new float[]{-1, -1, -1, -1, 9, -1, -1, -1, -1};
kernel.put(0, 0, data);
```

【例9-2】图像模糊与锐化的应用示例。
（1）创建应用程序框架，并设置应用程序的OpenCV环境。
（2）设计界面布局。
在界面中设置一个显示图片的ImageView组件和两个按钮组件，两个按钮分别用于模糊图像和锐化图像。
（3）设计主控制程序代码。

```
1   package com.example.ex9_2;
2   import androidx.appcompat.app.AppCompatActivity;
3   import android.os.Bundle;
4   import android.graphics.Bitmap;
5   import android.graphics.BitmapFactory;
6   import android.util.Log;
7   import android.view.View;
8   import android.widget.Button;
9   import android.widget.ImageView;
10  import org.opencv.android.OpenCVLoader;
11  import org.opencv.android.Utils;
12  import org.opencv.core.CvType;
```

```java
13      import org.opencv.core.Mat;
14      import org.opencv.core.Point;
15      import org.opencv.core.Size;
16      import org.opencv.imgproc.Imgproc;
17      public class MainActivity extends AppCompatActivity {
18          ImageView imgView;
19          Button btn1, btn2;
20          @Override
21          protected void onCreate(Bundle savedInstanceState) {
22              super.onCreate(savedInstanceState);
23              setContentView(R.layout.activity_main);
24              imgView = findViewById(R.id.imageView);
25              btn1 = findViewById(R.id.button);
26              btn2 = findViewById(R.id.button2);
27              btn1.setOnClickListener(new mClick1());
28              btn2.setOnClickListener(new mClick2());
29              boolean success = OpenCVLoader.initDebug();
30          }
31          class mClick1 implements View.OnClickListener {
32              @Override
33              public void onClick(View v) {
34                  Bitmap bitmap = BitmapFactory.decodeResource(
35                          MainActivity.this.getResources(), R.drawable.flower2);
36                  //定义3×3模糊卷积核
37                  Mat kernel = new Mat(3, 3, CvType.CV_32FC1);
38                  float[] data = new float[]{
39                                  0,          1.0f/8.0f,          0,
40                          1.0f/8.0f,          0.5f,       1.0f/8.0f,
41                                  0,          1.0f/8.0f,          0
42                  };
43                  kernel.put(0, 0, data);
44                  //使用上面定义的卷积核实现滤波
45                  Mat src = new Mat();
46                  Mat dst = new Mat();
47                  Utils.bitmapToMat(bitmap, src);
48                  Imgproc.filter2D(src, dst, -1, kernel, new Point(-1, -1),
49                          0.0, Imgproc.CCL_BOLELLI);
50                  //高斯模糊
51                  Imgproc.GaussianBlur(src, dst, new Size(5.0, 5.0), 10.0, 20.0);
52                  Utils.matToBitmap(dst, bitmap);
53                  imgView.setImageBitmap(bitmap);
54                  src.release();
55                  dst.release();
56              }
57          }
58          class mClick2 implements View.OnClickListener{
```

第9章 OpenCV应用实战——人脸美颜与人脸检测

```
59        @Override
60        public void onClick(View v) {
61          Bitmap bitmap = BitmapFactory.decodeResource(
62                 MainActivity.this.getResources(), R.drawable.flower2);
63          //定义锐化算子卷积核
64          Mat kernel = new Mat(3, 3, CvType.CV_32FC1);
65          float[] data = new float[]{-1, -1, -1, -1, 9, -1, -1, -1, -1};
66          kernel.put(0, 0, data);
67          //使用上面定义的卷积核实现滤波
68          Mat src = new Mat();
69          Mat dst = new Mat();
70          Utils.bitmapToMat(bitmap, src);
71          Imgproc.filter2D(src, dst, -1, kernel, new Point(-1, -1),
72                      0.0, Imgproc.CCL_BOLELLI);
73          Utils.matToBitmap(dst, bitmap);
74          imgView.setImageBitmap(bitmap);
75          src.release();
76          dst.release();
77        }
78     }
79  }
```

程序运行结果如图 9.10 所示。

图 9.10 图像模糊和图像锐化示例

9.2 人脸美颜

应用 OpenCV 进行人脸美颜，实现方法比较简单，主要实现方案是先结合高斯模糊与双边滤波进行磨皮，再调亮度，最后进行图像增强（锐化）处理。

下面通过示例介绍应用 OpenCV 对人脸进行美颜处理的设计方法。

【例 9-3】 照片的美颜处理示例。

（1）设置应用程序的 OpenCV 环境。

① 创建应用程序框架，并导入 OpenCV-android-sdk 框架下的 sdk\java 模块。

② 建立应用程序 App 与 OpenCV-android-sdk 的模块依赖关系。

③ 把 OpenCV 模块的编译版本 Compile Sdk Version 设置成与 App 的编译版本一致。

④ 在 Project 视图模式下，将 OpenCV-android-sdk 框架下的 sdk\native\libs 目录下所有文件复制到应用程序新建的 app\src\main\jniLibs 目录下。

⑤ 修改 app 目录下的 build.gradle 文件，在 android 节点中添加 jniLibs。

```
sourceSets {
    main {
        jniLibs.srcDirs = ['src/main/jniLibs']
    }
}
```

至此，OpenCV 的开发环境搭建完毕。

（2）设计界面布局。

在界面中设置一个显示图片的 ImageView 组件和一个进行美颜处理的按钮组件。

（3）设计主控制程序代码。

这里使用了美肤-磨皮算法，即先对图像进行双边滤波处理之后，再进行高斯模糊处理，最后调节图像的亮度，达到去除皮肤斑点和美白的效果，如图 9.11 所示。

（美颜前，脸部有很多斑痕）

（美颜后，经磨皮美颜处理）

图 9.11　照片的人脸美颜处理

完整的主控制程序代码如下。

```
1    package com.example.ex9_3;
2    import androidx.appcompat.app.AppCompatActivity;
3    import android.os.Bundle;
4    import android.graphics.Bitmap;
5    import android.graphics.BitmapFactory;
6    import android.os.AsyncTask;
7    import android.util.Log;
```

第9章　OpenCV应用实战——人脸美颜与人脸检测

```
8    import android.view.View;
9    import android.widget.Button;
10   import android.widget.ImageView;
11   import org.opencv.android.OpenCVLoader;
12   import org.opencv.android.Utils;
13   import org.opencv.core.Core;
14   import org.opencv.core.Mat;
15   import org.opencv.core.Scalar;
16   import org.opencv.core.Size;
17   import org.opencv.imgproc.Imgproc;

18   public class MainActivity extends AppCompatActivity {
19     private ImageView imageView;
20     private Bitmap processbp;
21     private Button btn;
22     private BilateralFilterTask bilateralFilterTask;

23     @Override
24     public void onCreate(Bundle savedInstanceState) {
25       super.onCreate(savedInstanceState);
26       setContentView(R.layout.activity_main);
27       btn = (Button)findViewById(R.id.button);
28       imageView = (ImageView)findViewById(R.id.imageView);
29       imageView.setScaleType(ImageView.ScaleType.FIT_CENTER);
30       btn.setOnClickListener(new mClick());
31     }
32     private class mClick implements View.OnClickListener{
33       @Override
34       public void onClick(View v) {
35         staticLoadCVLibraries();
36       }
37     }

38     //OpenCV库静态加载并初始化
39     private void staticLoadCVLibraries(){
40       boolean load = OpenCVLoader.initDebug();
41       convertGray(10);
42       if(load) {
43         Log.i("CV", "OpenCV Libraries loaded...");
44       }
45     }

46     private void convertGray(float bilityTraversal) {
47       bilateralFilterTask = new BilateralFilterTask();
48       bilateralFilterTask.execute(bilityTraversal);
49     }
```

```
50    private class BilateralFilterTask extends AsyncTask<Float, Bitmap, Bitmap>{
51      @Override
52      protected void onPreExecute() {
53        super.onPreExecute();
54      }
55      @Override
56      protected void onPostExecute(Bitmap bitmap) {
57          imageView.setImageBitmap(bitmap);
58      }
59      // 美肤-磨皮算法
60      protected Bitmap doInBackground(Float... bilityTraversal) {
61        processbp = BitmapFactory.decodeResource(
62                  MainActivity.this.getResources(), R.drawable.a2);
63        if (bilityTraversal[0] > 0) {
64          int dx = (int)bilityTraversal[0].floatValue()* 5; //设置双边滤波参数
65          double fc = bilityTraversal[0] * 12.5;  //设置另一个双边滤波参数
66          double p = 0.1f;      // 设置透明度
67          Mat image = new Mat(),
68              dst = new Mat(),
69              matBilFilter = new Mat(),
70              matGaussSrc = new Mat(),
71              matGaussDest = new Mat(),
72              matTmpDest = new Mat(),
73              matSubDest = new Mat(),
74              matTmpSrc = new Mat();
75          // 双边滤波
76          Utils.bitmapToMat(processbp, image);
77          Imgproc.cvtColor(image, image, Imgproc.COLOR_BGRA2BGR);
78          Imgproc.bilateralFilter(image, matBilFilter, dx, fc, fc);
79          Core.subtract(matBilFilter, image, matSubDest);
80          Core.add(matSubDest, new Scalar(128, 128, 128, 128), matGaussSrc);
81          // 高斯模糊
82          Imgproc.GaussianBlur(matGaussSrc, matGaussDest,
83          new Size(2 * bilityTraversal[0] - 1, 2 * bilityTraversal[0] - 1),
84                  0, 0);
85          matGaussDest.convertTo(matTmpSrc, matGaussDest.type(), 2, -255);
86          Core.add(image, matTmpSrc, matTmpDest);
87          Core.addWeighted(image, p, matTmpDest, 1 - p, 0.0, dst);//图像融合
88          Core.add(dst, new Scalar(10, 10, 10), dst);//提升亮度
89          Utils.matToBitmap(dst, processbp);//转换成 Bitmap 类型
90        }
91        return processbp;
92      }
93    }
94  }
```

第9章 OpenCV应用实战——人脸美颜与人脸检测

9.3 人脸检测

人脸检测定位，指的是对图片中的人脸进行检测，并给出每张人脸的定位和特征点信息。这样，人脸区域就可以被裁剪出来，经过预处理可以进一步进行人脸识别处理。这里介绍基于 OpenCV 开源的人脸特征库来实现人脸检测定位。

【例 9-4】 人脸检测示例。

（1）搭建 OpenCV 的开发环境。

（2）复制 OpenCV 的人脸特征文件。

在 OpenCV-android-sdk\sdk\etc\lbpcascades 目录下，找到 OpenCV 的人脸特征文件 lbpcascade_frontalface.xml，将其复制到应用程序中新建的 res\raw 目录下，如图 9.12 所示。

图 9.12 复制 OpenCV 的人脸特征文件

（3）设计界面布局。

在界面中设置一个显示图片的 ImageView 组件和一个检测人脸的按钮组件。

（4）设计主控制程序代码。

```
1   package com.example.ex9_4;
2   import androidx.appcompat.app.AppCompatActivity;
3   import android.os.Bundle;
4   import android.content.Context;
5   import android.graphics.Bitmap;
6   import android.graphics.BitmapFactory;
7   import android.util.Log;
8   import android.view.View;
9   import android.widget.Button;
10  import android.widget.ImageView;
11  import org.opencv.android.OpenCVLoader;
12  import org.opencv.android.Utils;
13  import org.opencv.core.Mat;
14  import org.opencv.core.MatOfRect;
15  import org.opencv.core.Rect;
```

```java
16  import org.opencv.core.Scalar;
17  import org.opencv.imgproc.Imgproc;
18  import org.opencv.objdetect.CascadeClassifier;
19  import java.io.File;
20  import java.io.FileOutputStream;
21  import java.io.InputStream;

22  public class MainActivity extends AppCompatActivity{
23  Button btn;
24  Mat src, dst, mGray, mRgba;
25  Mat ccs;
26  Bitmap bitmap, bitmap3;
27  ImageView img;
28  int mAbsoluteFaceSize;
29  CascadeClassifier classifier;
30  File cascadeFile;
31  @Override
32  protected void onCreate(Bundle savedInstanceState) {
33      super.onCreate(savedInstanceState);
34      setContentView(R.layout.activity_main);
35      img = (ImageView)findViewById(R.id.imageView);
36      btn = (Button) findViewById(R.id.button);
37      btn.setOnClickListener(new mClick());
38  }

39  class mClick implements View.OnClickListener{
40      public void onClick(View v) {
41        boolean success = OpenCVLoader.initDebug();
42        if(success)  initClassifier();
43        else Log.i("openCV_err", " OpenCVLoader.initDebug() err ......");
44      }
45  }
46  // 初始化人脸级联分类器
47  private void initClassifier() {
48   try {
49       //加载人脸级联分类器
50       InputStream inputStream = getResources()
51               .openRawResource(R.raw.lbpcascade_frontalface);
52       File cascadeDir = getDir("cascade", Context.MODE_PRIVATE);
53       cascadeFile = new File(cascadeDir, "lbpcascade_frontalface.xml");
54       FileOutputStream outputStream = new FileOutputStream(cascadeFile);
55       byte[] buffer = new byte[4096];
56       int bytesRead;
57       while ((bytesRead = inputStream.read(buffer)) != -1) {
58           outputStream.write(buffer, 0, bytesRead);
59       }
60       inputStream.close();
```

第9章 OpenCV应用实战——人脸美颜与人脸检测

```
61        outputStream.close();
62        classifier= new CascadeClassifier(cascadeFile.getAbsolutePath());
63        Log.i("classifier", "classifier = =" + classifier);
64        cascadeFile.delete();
65        cascadeDir.delete();
66        //读取人脸图片
67        bitmap = BitmapFactory.decodeResource(
68                MainActivity.this.getResources(), R.drawable.jiemei);
69        //生成灰度图,提高检测效率
70        setgray(bitmap);
71        ccs = onCamera(bitmap);
72        Utils.matToBitmap(ccs, bitmap);
73        img.setImageBitmap(bitmap);
74    }catch (Exception e) { e.printStackTrace(); }
75   }
76   //执行人脸检测的逻辑
77   Mat onCamera(Bitmap  inputFrame){
78     mGray = new Mat();
79     mRgba = new Mat();
80     Utils.bitmapToMat(inputFrame, mGray);
81     Utils.bitmapToMat(inputFrame, mRgba);
82     float mRelativeFaceSize = 0.8f;
83     if (mAbsoluteFaceSize == 0) {
84       int height = mGray.rows();
85       Log.i("Size", "mGray.rows() == " + height);
86       if (Math.round(height * mRelativeFaceSize) > 0) {
87          mAbsoluteFaceSize = Math.round(height * mRelativeFaceSize);
88       }
89     }
90     MatOfRect faces = new MatOfRect(dst);
91     if (classifier != null)
92       classifier.detectMultiScale(mRgba, faces);
93     //画矩形
94     Rect[] facesArray = faces.toArray();
95     Log.i("Size", "facesArray.length == " + facesArray.length);
96     Scalar faceRectColor = new Scalar(255, 255, 0, 255);
97     for(int i = 0; i < facesArray.length; i++)
98         Imgproc.rectangle(mRgba, facesArray[i].tl(),
99                     facesArray[i].br(),faceRectColor,12);
100    return mRgba;
101   }
102   // 生成灰度图片
103   void setgray(Bitmap bitmap2){
104    src = new Mat();
105    dst = new Mat();
106    Utils.bitmapToMat(bitmap2, src);
107    // Imgproc.cvtColor(src, dst, Imgproc.COLOR_BGRA2GRAY);
```

```
108        src.copyTo(dst);
109        Utils.matToBitmap(dst, bitmap2);
110        Log.i("Size", "dst == " + dst.toString());
111        src.release();
112        dst.release();
113    }
114 }
```

程序运行结果如图 9.13 所示。

图 9.13　人脸检测

附表　微课视频二维码索引列表

序号	视频内容标题		视频二维码位置	所在页码
1	第 1 次课（1）	Android Studio 的下载与安装	1.2 搭建 Android Studio 开发环境	4
2	第 1 次课（2）（例 1-1）	Android 应用程序设计示例	1.6 Android 应用程序设计示例（例 1-1）	19
3	第 1 次课（3）（例 1-2）	Android 应用程序设计示例	1.6 Android 应用程序设计示例（例 1-2）	20
4	第 2 次课（1）	Android 项目结构	1.5 Android 项目结构	12
5	第 2 次课（2）	设置单行文本内容	2.3.1 文本标签	33
6	第 2 次课（3）	设置多行文本内容	2.3.1 文本标签	33
7	第 3 次课（1）	文本组件	2.4 文本编辑框 EditText	38
8	第 3 次课（2）	按钮	2.3.2 按钮及按钮处理事件	36
9	第 4 次课（1）	布局设计_基本布局	2.2 Android 布局管理	22
10	第 4 次课（2）	布局设计_图片浏览器	2.6 图像显示 ImageView	51
11	第 5 次课（1）	进度条	2.5.1 进度条	42
12	第 5 次课（2）	多页面切换	3.1.1 绑定机制组件 Intent 3.1.2 Activity 页面切换	64
13	第 6 次课　页面之间传递数据		3.1.3 在 Activity 页面之间传递数据	68
14	第 7 次课　图形绘制		4.1.1 几何图形绘制类 4.1.2 几何图形绘制过程	94
15	第 8 次课（1）	自定义组件	4.1.3 自定义组件	97
16	第 8 次课（2）	触摸屏事件	4.2.1 简单的触摸屏事件	100
17	第 9 次课（1）	手势划屏	4.2.2 手势识别（例 4-4）	102
18	第 9 次课（2）	画图板	4.2.2 手势识别（例 4-5）	104
19	第 10 次课（1）	音乐播放器设计	4.3.3 播放音频文件（例 4-6）	111
20	第 10 次课（2）	带进度指示条的音乐播放器	4.3.3 播放音频文件（例 4-7）	112
21	第 11 次课　视频播放器设计		4.4 视频播放（例 4-8）	117
22	第 12 次课　Fragment 组件_底部导航		3.4 Fragment	88
23	第 13 次课　Fragment 组件_侧部导航		3.4 Fragment（内容扩展）	88
24	第 14 次课　在 Fragment 的 Java 程序中添加组件		3.4 Fragment（内容扩展）	91
25	第 15 次课　推箱子游戏设计		4.6 动画技术（内容扩展）	126
26	第 16 次课　人脸美颜基 OpenCV_环境的建立		9.1.1 搭建 OpenCV Android 开发环境	237

序号	视频内容标题	视频二维码位置	所在页码
27	第16次课 人脸美颜基OpenCV_彩色转灰度	9.1.2 Mat对象与Bitmap对象	240
28	第17次课 人脸美颜基OpenCV_模糊与锐化	9.1.3 图像的模糊与锐化	242
29	第18次课（1） 人脸美颜	9.2 人脸美颜	245
30	第18次课（2） 人脸检测	9.3 人脸检测	249
31	第19次课（1） SQLite数据库_创建	8.2.3 SQLite数据库的操作命令（例8-1）	215
32	第19次课（2） SQLite数据库_使用	8.2.3 SQLite数据库的操作命令（例8-2及内容扩展）	219
33	第20次课（1） Volley框架	7.1 Volley框架及其应用	185
34	第20次课（2） 读取远程数据	7.1.4 Volley的基本使用方法（例7-2及内容扩展）	192
35	第21次课（1） 把数据写入后台服务器	7.3.1 把数据写入远程数据库	202
36	第21次课（2） 把数据写入后台数据库	7.3.1 把数据写入远程数据库	202
37	第22次课 读取远程数据库数据	7.3.2 读取远程数据库数据	205

图书资源支持

感谢您一直以来对清华版图书的支持和爱护。为了配合本书的使用,本书提供配套的资源,有需求的读者请扫描下方的"书圈"微信公众号二维码,在图书专区下载,也可以拨打电话或发送电子邮件咨询。

如果您在使用本书的过程中遇到了什么问题,或者有相关图书出版计划,也请您发邮件告诉我们,以便我们更好地为您服务。

我们的联系方式:

地 址:北京市海淀区双清路学研大厦 A 座 714

邮 编:100084

电 话:010-83470236 010-83470237

客服邮箱:2301891038@qq.com

QQ:2301891038(请写明您的单位和姓名)

资源下载: 关注公众号"书圈"下载配套资源。

书圈

清华计算机学堂

观看课程直播